Field Guide to

WISCONSIN SEDGES

This book was published with the support of the Arboretum at the University of Wisconsin–Madison, the Bradshaw-Knight Foundation, the Walter Kuhlman Award of the Diversity Inventory Group, the Botanical Club of Wisconsin, and the Citizens Natural Resources Association of Wisconsin, Inc.

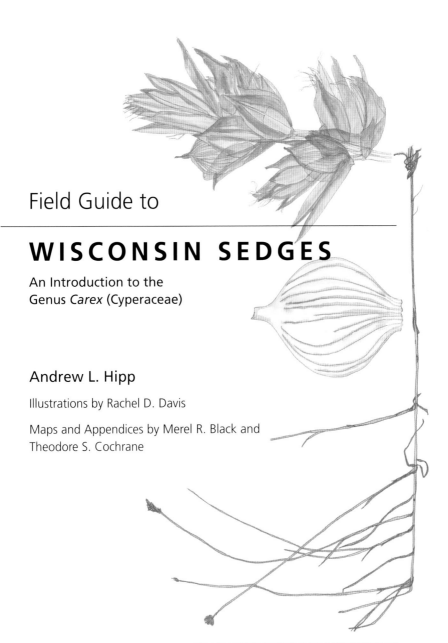

Field Guide to

WISCONSIN SEDGES

An Introduction to the
Genus *Carex* (Cyperaceae)

Andrew L. Hipp

Illustrations by Rachel D. Davis

Maps and Appendices by Merel R. Black and
Theodore S. Cochrane

THE UNIVERSITY OF WISCONSIN PRESS

Published in collaboration with the University of Wisconsin Arboretum and the
University of Wisconsin–Madison Department of Botany

The University of Wisconsin Press
1930 Monroe Street, 3rd Floor
Madison, Wisconsin 53711-2059

www.wisc.edu/wisconsinpress/

3 Henrietta Street
London WC2E 8LU, England

5   4   3   2   1

Printed in the United States of America

Library of Congress Cataloging-in-Publication Data
Hipp, Andrew.
   Field guide to Wisconsin sedges : an introduction to the
   genus Carex (Cyperaceae) / Andrew L. Hipp; illustrations
   by Rachel D. Davis.
       p. cm.
   Published in collaboration with the University of Wisconsin
   Arboretum and the University of Wisconsin—Madison
   Department of Botany.
   Includes bibliographical references and index.
   ISBN 0-299-22590-9 (cloth : alk. paper)—
   ISBN 0-299-22594-1 (pbk. : alk. paper)
       1. Carex—Wisconsin. I. Title.
QK495.C997H57   2008
584´.8409775—dc22   2007042153

For our sons,

DAVID and **LOUIS**,

born while this book was still in the works

Contents

Preface and Acknowledgments

This book is a guide to the identification of Wisconsin's *Carex* species, written for naturalists at varying levels of taxonomic expertise. Part 1 provides dichotomous keys to and brief descriptions of the 157 *Carex* species that inhabit the state. Where practical, keys and descriptions emphasize those characters that are easiest to evaluate in the field, deemphasizing measurements when reliable qualitative characters are available. Part 1 is organized according to the sectional classification followed in *Flora of North America* (*FNA* hereafter in the book). Thus, learning sedges using this book will provide experience useful to learning sedges anywhere else in the country.

Part 2 presents expanded descriptions and illustrations for approximately half of the state's species in a less technical, field guide format. The book also introduces the taxonomy and vocabulary needed to understand the genus. Because many of Wisconsin's sedges are common throughout much of the upper Midwest, the guide is useful as an introduction to the genus in adjacent states.

This book should serve as a bridge between field guides, which generally include a small number of sedges (if any) and provide little information about classification but offer good illustrations and guidelines for identifying common species, and technical manuals, which are needed to make accurate identifications but present little introduction to terminology. Two appendices provide information on Wisconsin *Carex* habitats, based on observation of Wisconsin specimens in the herbarium and field (by Theodore S. Cochrane), and range distributions in the state, based on herbarium specimens (maps by Merel R. Black and Theodore S. Cochrane). These, combined with the identification and habitat information in parts 1 and 2, make the book especially relevant to ecologists, land managers, naturalists, and native plant enthusiasts. We hope that the book will contribute to the substantial increase seen in recent years in the incorporation of sedges into ecological restorations.

   This book could not have been written without help from many individuals. I am especially grateful to Mark A. Wetter, Theodore Cochrane, and Paul E. Berry for providing space and herbarium resources for this project and to the University of Wisconsin–Madison Arboretum for encouraging this project and providing financial support for illustrations and publication. Molly Fifield Murray encouraged me greatly in my study of sedges while I was still a ranger at the UW–Madison Arboretum, and a conversation we had during that time was a significant impetus behind this book. Additional funding for the illustrations and publication costs was provided by the Citizens Natural Resources Association of Wisconsin, Inc., the Diversity Inventory Group, the Botanical Club of Wisconsin, and the Knight Foundation. Numerous conversations and field trips with Theodore Cochrane, Anton A. Reznicek, David A. Rogers, Paul E. Rothrock, and Elizabeth H. Zimmerman played an essential role in my understanding of the Wisconsin sedge flora. Sections of the manuscript were critically reviewed by Marlin Bowles, Theodore Cochrane, William J. Crins, David J. Egan, John L. Larson, David F. Murray, Anton A. Reznicek, and Paul E. Rothrock. Alison Mahoney workshopped the keys and text during an intensive week in the herbarium at the Morton Arboretum, and many of the most readable portions of the keys are a product of her acumen and refusal to let sleeping dogs lie. Marvin Jay Colbert carefully proofread many pages, and Jason Sturner and Jaime Weber assisted with indexing. Kathy Miner digitized the illustrations with utmost care; we would have been hesitant to entrust them to anyone else. Most of this book was written while I held an O. N. Allen Fellowship through the UW–Madison Department of Botany and, subsequently, in my current position at the Morton Arboretum. The time afforded me by these positions was invaluable to the completion of the project.

   Above all, I am grateful to my family. Rachel Davis, my wife, friend, and collaborator, was a joy to work with on this project, and our sons, David and Louis, were a continual joy and inspiration. Rachel and I are blessed with the support and love of family in all our endeavors. We cannot acknowledge them enough.

Abbreviations

FNA Ball, P. W., and A. A. Reznicek, 2002: *Flora of North America,* vol. 23: *Magnoliophyta: Commelinidae (in part): Cyperaceae,* 254–572.

<   less than

>   more than

≤   less than or equal to

≥   more than or equal to

±   more or less, approximately

f.   forma

sp.   species (singular)

spp.   species (plural)

ssp.   subspecies

var.   variety

Standard metric abbreviations are used for measurements.

Introduction to Sedges and
Use of This Book

Introduction

The sedge family, Cyperaceae, numbers approximately 5,000 species world-wide. Its largest genus, *Carex*, contains about 2,000 species. *Carex* is one of the world's largest flowering plant genera and one of the most widespread in a range of temperate ecosystems. Sedges grow in alpine tundra and bottomland forests, dry prairies, sedge meadows, peat lands of all kinds, upland forests, ditches and roadsides, and many other habitats. Sedges form an important component of forest understories, provide food for waterfowl and habitat for invertebrates, and carry fire through wetlands, prairies, and oak woodlands. Sedges directly increase the biodiversity of some communities: the widespread wetland tussock sedge (*Carex stricta*), for example, creates tall, peaty tussocks that provide suitable microhabitats for a great diversity of species. Moreover, *Carex* species tend to track habitats closely, making sedges useful habitat indicators. An understanding of the genus is critical to understanding the ecology of temperate North American plant communities.

Knowing sedges has practical benefits to the ecological restorationist and land manager as well. Although *Carex* forms an important component of wet prairies and sedge meadows, hydrological restoration of degraded wetlands is not sufficient to restore sedge diversity. In fact, sedges are among the last species to return to wetlands without active efforts to reintroduce them, apparently because of limited seed dispersal and persistence in the seed bank. Approximately a third of the *Carex* species of the upper Midwest are available as seed or plants from commercial native plant nurseries, and the habitat guidelines in this book provide information useful in deciding where to plant different sedge species. Publications by Kettenring (2006), Schütz (2000), van der Valk (1999), and their colleagues provide concrete guidelines for propagating *Carex* from seeds and, to a lesser extent, rhizomes. Using the identification skills you develop with this field guide, your choice of sedges for restoration projects need not be limited by commercial seed and plant sources.

The diversity and ecological importance of sedges was not the starting place for this book, however; rather, it was their beauty. Thumb through the illustrations in part 2, the field guide portion of this book, and you will find a surprising range of morphological diversity. Sedges exhibit various and fascinating variations on a simple theme. Admittedly, there are constraints on the evolution of the sedge body. No sedge will ever produce a flower of the complexity of an orchid, and there will never be finely dissected leaves or big fleshy fruits among the sedges. Yet within their constraints you will find a wide range of colors, textures, shapes, and growth forms.

Whatever your interest in sedges, this book will provide you the information you need to identify and understand Wisconsin sedges. We hope you find as much pleasure in your studies as we have in ours.

USING THIS BOOK

If you are not already familiar with sedges, take time to read the introductory chapters on sedge morphology, sedge taxonomy, and field study. Then familiarize yourself with some of the terminology and structures by studying some of the illustrations in part 2 and the species descriptions that accompany them. Use the glossary to look up terms you don't know. Once you have become a bit more familiar with the sedge body, the keys in part 1 will be easier to use. Learning sedges with this book will also help make the more technical resources listed in the bibliography accessible. Use this book as an introduction to sedges and their literature, a springboard to identifying Midwest sedges and understanding their ecology.

Part 1: Keys and Abbreviated Descriptions of Wisconsin Carex Sections and Species

Keys are provided for all species in Wisconsin, emphasizing characters likely to be of particular use in the field. Measurements are mostly rounded to the nearest 0.5 mm except where more precise measurements are needed to make accurate identifications. Although it is difficult to identify sedges reliably without making some measurements, a preoccupation with measuring can become a stumbling block to learning the genus. The keys and descriptions are intended to provide enough information to identify species, but they are simplified relative to the treatments in *FNA* and Fernald (1950). Characters are selected based on my impressions of what makes for easiest and most reliable identification in the field. For the sake of clarity, most taxa key out in only one place in this key. Morphologically aberrant individuals may key out incorrectly. It is advisable to study several plants in a population as you work through the keys and to use whole-plant material with ripe perigynia.

This section of the book also presents an up-to-date overview of sedge taxonomy that with few exceptions follows *FNA*. It includes brief descriptions of sections and their habitats as well as characterization of groups of sections that are morphologically allied to one another and may be easily confused. The keys and species descriptions rely on sectional characters insofar as it is not cumbersome to do so, so that learning Wisconsin's sedges with this book should help you learn the major sections of sedges that occur throughout North America. Characters that apply to all members of the section are typically not mentioned in the species descriptions. Consequently, you will need to fill in the gaps in each species description based on the section description that precedes it.

Part 2: Field Guide to Wisconsin Carices

The second half of this book serves as a field guide to Wisconsin's common *Carex* species. Roughly 80 percent of Wisconsin's carices, including nearly all of the species that you will encounter on a typical walk in the woods, are either illustrated in this section or discussed as or in conjunction with a look-alike. Each species description is accompanied by a habitat description and practical guidelines for discriminating between similar species. Like any field guide, this section of the book is only an introduction to the flora, but it may be a good place to start if the keys appear intimidating. Use the keys in part 1 to double-check your identifications. If you are not familiar with using technical botanical manuals, working iteratively between parts 1 and 2 of this book is a good way to learn the ins and outs of keying plants.

ABOUT THE ILLUSTRATIONS AND APPENDICES

The illustrations were executed in ink and watercolor and in most cases represent herbarium specimens housed at the University of Wisconsin–Madison Herbarium (WIS). A few plants were collected fresh for illustration. Each plant or plant part was illustrated separately, then digitized. I imported these digitized images into Adobe Photoshop and laid them out as plates. Most plates portray a whole plant, inflorescence, and perigynium for a single species, but many plates incorporate material from more than one taxon. Where this is the case, the caption and labels identify all parts illustrated.

The appendices are the work of Theodore S. Cochrane and Merel R. Black and reflect decades of caricological study by Mr. Cochrane.

SOURCES

Habitats and associated species reported in parts 1 and 2 are summarized from herbarium material housed at WIS and my experience of the plants in the field. Habitats in some cases do not generalize well to other states. Species descriptions are based on inspection of Wisconsin specimens and on published treatments, especially *FNA* and Fernald (1950); measurements are drawn largely from *FNA*, with modifications mostly where the measurement range in *FNA* is substantially broader than found in Wisconsin material. Keys are likewise based both on inspection of Wisconsin specimens and on published treatments. Any errors in the book are my own.

What Is a Sedge?

In common parlance, sedges and rushes often become just "some kind of grass." Yet sedges, grasses, and rushes, which make up almost all of the grasslike plants in temperate ecosystems, comprise three separate families: the Cyperaceae (the sedge family, approximately 5,000 species worldwide), Poaceae (the grass family, also known by the name Gramineae, approximately 10,000 species), and Juncaceae (the rush family, approximately 400 species). We need only look at a few common names to see the way in which these families are intermixed in the common view: spike-rushes (*Eleocharis* spp.), bulrushes (*Scirpus* spp., *Schoenoplectus* spp.), nut-rushes (*Scleria* spp.), and even some common "wiregrasses" (e.g., *Carex lasiocarpa* and *C. oligosperma*) are members of the Cyperaceae, not true rushes or grasses.

These families, while closely related to one another, do not interbreed and are ecologically and morphologically distinct from one another. Moreover, with a little study they are easy to distinguish from one another. Most of the major genera are easy to tell apart as well. By the end of this chapter you should be able to distinguish the Poaceae, Cyperaceae, and Juncaceae and recognize the genus *Carex* in the field. Once you can do this, you will be well on your way to learning sedges.

Distinguishing Sedges from Grasses and Rushes

Distinguishing the Cyperaceae, Juncaceae, and Poaceae based on inflorescence characters is straightforward with a hand lens and fruiting or flowering material. The inflorescence of the rushes (Juncaceae) is the simplest of the three families, with *flowers typically bisexual, subtended by 6 small, green or brown, petal-like structures (tepals)*. The rush flower thus resembles a small lily, and rushes were in the past hypothesized to be closely related to the lilies, with flowers reduced under selective pressure for wind pollination. Considerable anatomical and molecular data, however, demonstrate that in

6

fact the Juncaceae are sister to the Cyperaceae, and neither is very closely related to the lilies. The sedge (Cyperaceae) inflorescence comprises *unisexual or bisexual flowers that are each subtended by a single scale*. In *Carex*, discussed in greater detail below, each female flower is additionally enclosed in a highly modified bract called a perigynium or utricle. Grasses (Poaceae) are the most complex of the three families in inflorescence structure, producing *bisexual flowers surrounded by numerous specialized scales*. Each individual flower, or floret, is enclosed in a two-scale sandwich, a lemma on the outside and a palea on the inside. The florets are borne in inflorescence units called spikelets, which are themselves surrounded by specialized scales called glumes.

The families are also easily recognized by the structure of the fruits. As with their inflorescence structure, rushes (Juncaceae) are the most distinctive of the three families in fruit type, producing a capsule that contains many small seeds in the genus *Juncus* (the rushes), three seeds per capsule in *Luzula* (the wood rushes). The morphology of these seeds is key to identifying the different species of *Juncus*, making the collection of mature fruits as important in rush taxonomy as it is in identifying sedges. The sedges (Cyperaceae) produce small achenes that are highly variable in morphology. They range from lens-shaped (lenticular) to three-sided or round in cross-section. They may be black, chestnut, or green. Their surfaces range from entirely smooth to highly rugose, roughened, or striated, with prominent bristles characteristic of species in many genera. This diversity is restricted within the genus *Carex*, in which achenes are enclosed within the perigynium and typically smooth, almost never ornamented. Achene characteristics are nonetheless important in the identification of many species in the genus. The grasses (Poaceae) produce a *specialized achene called a karyopsis*, which differs in having the seed coat completely fused to the fruit wall.

Yet often only vegetative structures are available, so it is fortunate that we can distinguish the families by vegetative characters alone. Many people are first introduced to graminoid taxonomy through the rhyme "sedges have edges." In fact, the shoots of many sedges are triangular in cross-section. The genus *Carex* in particular has many species with prominently triangular culms and vegetative shoots. Many of the bulrushes (species in the genus *Schoenoplectus* and some of its allies) and spike-rushes (*Eleocharis*), however, have culms that are more often round in cross-section. Culm cross-section also provides useful characters for distinguishing the families: while the culms of grasses are typically hollow, with prominent, jointlike nodes, the culms of sedges and rushes are more often solid between the nodes and inconspicuously jointed. There are exceptions, of course, including the hollow-stemmed plants of genus *Dulichium* and *Carex* section *Ovales*.

Phyllotaxy—the distribution of leaves on a plant—often suffices to distinguish the Cyperaceae, in which the leaves are alternate and three-ranked, meaning that they arise as though off of the points of a triangle. The three-ranked condition is particularly obvious in three-way sedge (*Dulichium arundinaceum*) and Muskingum sedge (*Carex muskingumensis*) but evident in virtually all species in the family that have well-developed leaf blades. Leaves of rushes are also three-ranked, while those of grasses are alternate and two-ranked, arising from opposite sides of the culm. This is particularly obvious in the common wetland manna grasses (*Glyceria* spp.), which occur with many of our wetland sedges.

Leaves provide two additional characters that are useful for distinguishing the families from one another. The more obvious of these is the leaf sheath, which is a closed, tubelike structure in the sedges but open in the front in sedges and rushes. The appearance of the inner band of the leaf sheath, opposite the blade, is like that of a crewneck T-shirt in the Cyperaceae. The inner band of the leaf sheath in grasses and rushes is not fused and more nearly resembles the overlapping edges of a robe. Less obvious but at least as useful as the leaf sheath in distinguishing the families is the ligule, an extension of the leaf sheath that originates at the base of the leaf blade. In sedges the ligule is a flap of tissue, typically triangular, that is fused to the inner face of the leaf blade. The margins of the ligule are loose and membranous. In grasses the ligule is loose rather than fused to the leaf blade and highly variable, ranging from membranous to fibrous and long to inconspicuous. Ligule characters are often useful in identifying grasses. In rushes the ligule is free of the leaf blade but wraps around the stem, unlike either the sedges or the grasses.

RECOGNIZING THE GENUS *CAREX*

Once you know that the plant in your hand is in the sedge family Cyperaceae, identifying it as a member of the genus *Carex* is usually straightforward. In *Carex* the flowers are unisexual, and the female flower is surrounded by a saclike structure called the *perigynium* (*peri-* meaning "around," *-gyn-* referring to the female [pistillate] flowers; pronounce this word to rhyme with "condominium"). Inside the perigynium the ovary ripens to become an *achene*, which is either lens shaped or three sided. A style emerges from the apex of the achene and terminates in a *stigma* that is divided into two or three sections (rarely four). If you find a Wisconsin sedge with a perigynium, the species should be in this book.

Identifying species within *Carex* requires a familiarity with several aspects of the sedge body. In the text that follows, each paragraph begins with one line from the rather formal description of the genus *Carex* that is included in *FNA*, which is the most comprehensive and usable single source for the taxonomy of North American sedges. Throughout the text, words that are italicized appear in the glossary at the end of this book. By the end of this overview, you should sufficiently understand the genus description, the sedge body, and the vocabulary of sedge *morphology* to make sense of the keys and descriptions in this book. If you can, work through the description with a fresh *Carex* specimen at hand, moving from the base to the apex of the plant.

Herbs, perennial, cespitose or not, rhizomatous, rarely stoloniferous

Carex species in our region are all rhizomatous perennials, with wide variation in the *rhizome* length. Plants may be *cespitose*, forming clumps of shoots, typically with rhizomes very short and often inconspicuous; or shoots may arise singly from nodes of a creeping rhizome or *stolon*. Slender, creeping rhizomes play an important role in vegetative reproduction in many species. In *Carex stricta* (tussock sedge), for example, the rhizomes are of two types, one set spreading horizontally and functioning primarily in vegetative reproduction, the other growing vertically and aiding in the formation of tussocks. Some of our *Carex* species, notably in section *Ovales*, are short-lived and have the capacity to go through a generation in a single year. All, however, are perennials.

Culms usually trigonous, sometimes round

As in grasses and rushes, the sedge stem is referred to as a *culm*. The typically triangular cross-section of the culm is the source of the adage that "sedges have edges." While such sedge genera as *Eleocharis* and *Schoenoplectus* do not have trigonous culms, most *Carex* species do, though some have round or only obtusely angled culms. Related to this characteristic is the fact that leaves in *Carex* are tristichous, meaning that they grow in three rows or ranks. Looking down on a sedge shoot from the end, you should be able to see that the leaves arise in a spiral in three rows, as though off the points of a triangle.

In most species all culms bear inflorescences, and vegetative shoots are composed of overlapping sets of leaf sheaths: such vegetative shoots are sometimes referred to as *pseudoculms*. In some carices (e.g., the species of sections *Ovales, Carex,* and *Holarrhenae*) the vegetative shoots are true *vegetative culms.* The distinction between these types of vegetative shoots can easily be seen by making a cross-section about one-third of the way up the shoot. In a true vegetative culm the cross-section includes a solid or hollow stem; in a typical vegetative shoot the cross-section passes only through leaf sheaths.

Leaves basal and cauline, sometimes all basal; ligules present; blades flat, V-shaped, or M-shaped in cross-section, rarely filiform, involute, or rounded, commonly less than 20 mm wide, if flat then with distinct midvein

As in grasses, rushes, and such other monocots as cattails (*Typha*), the sedge leaf is divided into a blade and a sheath. The leaf blade is usually flat or folded lengthwise (V-shaped or M-shaped in cross-section). Leaves are typically very thin, but in some of our wetland species the blades and often the sheaths are thickened with air-filled tissue, or aerenchyma. Upon drying, the divisions, or *septa,* between these air pockets become apparent as the leaf tissue around them shrinks. Such leaves are referred to as *septate-nodulose.* The leaf blade connects to the back or dorsal side of the leaf sheath. Fused to the inner face of the leaf blade at the point of connection with the sheath is an upside-down V-shaped, U-shaped, or nearly squared-off (truncate) flap of tissue called the *ligule.* Whereas in grasses the entire ligule is free, only the margin of the ligule is free in sedges.

Leaf sheaths often bear crucial taxonomic characters. The front or ventral side of the sheath—the side opposite the leaf blade, referred to throughout this book as the inner band of the leaf sheath—may be green or *hyaline,* veined or not. It may be smooth, *pubescent, papillose,* or *corrugated.* Most of these features are easily seen with the naked eye, but a hand lens is typically necessary to detect the very tiny raised bumps on a papillose leaf sheath and the shortest types of pubescence. Some characters become obvious only as the leaf sheath begins to break down. Leaf sheaths of the common tussock sedge (*Carex stricta*), for example, break down to reveal an interlocking reticulum of veins typically described as *ladder-fibrillose* or pinnate-fibrillose. In other species, such as Pennsylvania sedge (*C. pensylvanica*) and Sprengel's sedge (*C. sprengelii*), the veins persist as a vertically aligned tuft or shroud of fibers at the base of the plant.

The lowest leaf sheaths may bear leaf blades, in which case the plant is referred to as *phyllopodic* (*phyllo-* meaning "leaf," *-podic* from the word for "foot"; thus, "phyllopodic" translates approximately as "leaf-footed"). Sedges with basal sheaths that lack leaf blades are referred to as *aphyllopodic.*

Basal sheaths are distinctly colored in many species: this alone is a good reason to make whole-plant collections when you are out sedging.

Inflorescences terminal, consisting of spikelets borne in spikes arranged in spikes, racemes, or panicles; bracts subtending spikes leaflike or scalelike; bracts subtending spikelets scalelike, very rarely leaflike. Spikelets 1-flowered; scales 0–1

"Spikelets borne in spikes arranged in spikes?!" The inflorescence in sedges can be difficult to interpret when you first look at it, though certainly not as difficult as the grass inflorescence, with its proliferation of specialized scales. The spikelet is the basic unit of the inflorescence: a single, unisexual flower with its associated single *scale*. In a very few cases, the "scale" of a sedge spikelet is in fact a leaflike *bract* (see *Carex jamesii* for a good example of this). Spikelets are *sessile* (stalkless), the spikelets borne on a single inflorescence axis forming a spike. These spikes are variously stalked or sessile, branching or simple, bisexual or unisexual, depending on the section and the species. In this book the term *inflorescence* is used for the entire collection of flowers and associated bracts, while *spike* is reserved for a collection of flowers on a single inflorescence axis. To bypass unnecessary confusion with the specialized terminology of grass taxonomy, I avoid the term *spikelet* in this book.

Flowers unisexual; staminate flowers without scales; pistillate flowers with 1 scale with fused margins (perigynium) enclosing flower, open only at apex; perianth absent; stamens 1–3; styles deciduous or rarely persistent, linear, 2–3 (–4)-fid.

Unlike most other sedge genera, the *Carex* flower is invariably unisexual. *Staminate* ("male") flowers are *subtended* by a single scale and are composed of 1–3 stamens. At anthesis (full flowering), the *anthers* may be bright yellow and conspicuous. The anthers fall off soon after pollen is dispersed, leaving only the slender filaments.

The *pistillate* ("female") flowers are surrounded by the *perigynium* (commonly referred to in European texts by the more general term *utricle*), a saclike structure with an opening at the apex through which the *stigma* emerges. Perigynia display modifications for flotation, adhesion, wind dispersal, ant dispersal, and ingestion. Many of the most important characters for species identification are on the perigynium, and measurements to the nearest 0.1 mm are sometimes required to make precise identifications. For the most part, measurements in this book are rounded to the nearest 0.5 mm, but more precise measurements are retained where they are crucial for distinguishing between similar species.

Achenes biconvex, plano-convex, or trigonous, rarely 4-angled

As indicated above, the fruit type alone is sufficient to distinguish the families Cyperaceae, Poaceae, and Juncaceae. Of these three families, only the Cyperaceae produce achenes. Relative to the rest of the family, *Carex* achenes show comparatively few distinguishing characters between species. While the bristles and surface texture of achenes provide numerous taxonomic characters in most genera, those in *Carex* are for the most part unremarkable. There are exceptions. A key difference between *Carex lupulina* and *C. lupuliformis,* for example, is the distinct knobbing on the achenes of the latter. Likewise, the lovely *C. tuckermanii* is marked by achenes that are invaginated on one edge, as though the achene had been pinched during development. But for the most part the perigynium appears to have shielded the achene surface from natural selection, so that while the perigynium has adapted to different modes of dispersal, achene morphology is less variable in *Carex* than in the other sedge genera.

Achenes are triangular in cross-section (trigonous) if the ovaries from which they derive are made of three fused carpels, lens shaped (lenticular) if the ovaries are made of two carpels. This difference in shape is of course best seen if the perigynium is split open, but often it can be inferred from the cross-sectional shape of the perigynium itself. Because the number of stigmatic lobes reflects the number of fused carpels, achene cross-sectional shape can also be deduced from the stigma morphology, though stigmas are often damaged later in the season and are absent from many specimens.

x = 10

Finally, we come to the chromosomes, which are among the most interesting aspects of sedge biology. While most chromosomes (e.g., those of grasses and humans) attach to the mitotic spindle by a single, localized centromere, sedges and rushes have holocentric chromosomes, which behave as though they had many centromeres distributed along the chromosome arms. Such chromosomes have the capacity to evolve rapidly by fission (chromosome breakage) and fusion. This form of chromosome evolution is known as agmatoploidy. The potential for rapid chromosome evolution very likely plays an important role in the evolution and diversification of sedges.

In most plant groups, chromosome evolution proceeds primarily through duplication of the entire set of chromosomes. If such were the case in *Carex,* the equation "$x = 10$" would tell us that the most primitive condition in the genus was likely a haploid chromosome count of 10. In *Carex*, however, this concept of base chromosome number is difficult to interpret, because chromosome arrangements in the genus shift so rapidly within and between species that the base chromosome number probably cannot be inferred with any confidence. For purposes of basic sedge taxonomy, it suffices to know that chromosome numbers often provide useful taxonomic information for distinguishing closely related *Carex* species. Still, much remains to be learned about patterns and processes of chromosome evolution and the role of chromosomal rearrangements in the formation of new sedge species.

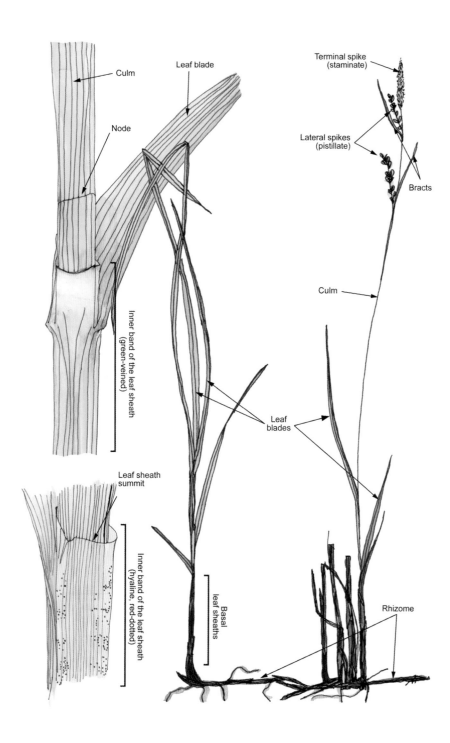

Culm

Leaf blade

Terminal spike
(staminate)

Node

Lateral spikes
(pistillate)

Bracts

Inner band of the leaf sheath
(green-veined)

Culm

Leaf sheath
summit

Leaf
blades

Inner band of the leaf sheath
(hyaline, red-dotted)

Basal
leaf sheaths

Rhizome

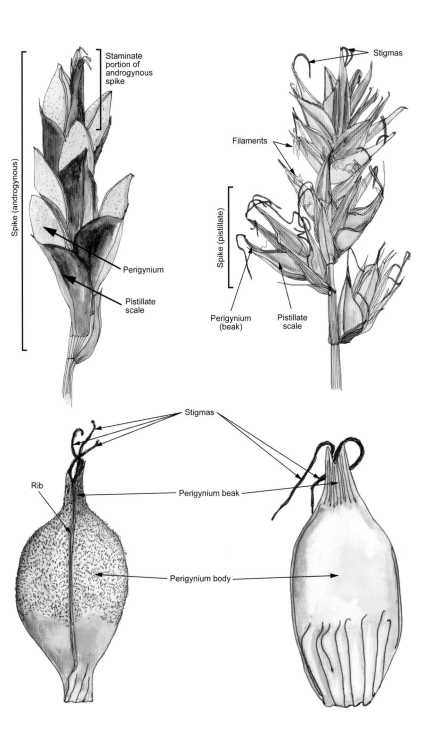

Staminate portion of androgynous spike

Spike (androgynous)

Perigynium

Pistillate scale

Stigmas

Filaments

Spike (pistillate)

Perigynium (beak)

Pistillate scale

Stigmas

Rib

Perigynium beak

Perigynium body

Sedge Taxonomy

> Taxonomy might be variously defined, according to the definer, but I believe the best
> definition is "a study aimed at producing a system of classification or organisms
> which best reflects the totality of their similarities and differences." This is obviously a
> large task.
>
> —ARTHUR CRONQUIST, *The Evolution and Classification of Flowering Plants*

The field of taxonomy encompasses several interrelated areas of study: nomenclature, the system of scientific (Latin) names applied according to a set of standard, internationally recognized rules; classification, the system by which species are gathered into an increasingly complex hierarchy of groupings, including genera, families, orders, etc.; species circumscription, the description of what (usually wild) populations constitute the functional groups that we refer to as species; and phylogenetics, the study of species relationships and character evolution. Ultimately, these studies are basic to almost all questions regarding ecology and evolution. The genus *Carex*, as one of the world's largest and most challenging flowering plant genera, is rife with taxonomic questions. Of these, nomenclature and classification pertain directly to the concerns of this book.

NOMENCLATURE

Common names of plants are often deceiving. Grasses, rushes, and sedges are no exception. All of the approximately 5,000 species in the family Cyperaceae are properly referred to as sedges. Yet within the Cyperaceae are plants referred to as "spike-rushes," "nut-rushes," "wiregrasses," "wool-grasses," and "umbrella-grasses."

Common names in *Carex* tend to be dull. Common wood sedge. Prairie gray sedge. Bristle-leaf sedge. There is a reason for this monotony: most common names in *Carex* have been applied by botanists for practical purposes, often based on direct translations of the more sonorous Latin names. In contrast, many North American grass names are evocative of people's direct experiences with the plants. Redtop. Bluejoint. Turkey-foot. Foxtail. Sprangletop. Sandbur. Dropseed. This likely has much to do with the fact that grasses are important forage and food plants; consequently, they are known more generally. Sedges, on the other hand, are not usually as competitive as grasses in uplands and are less important as forage plants; consequently, they are more poorly known outside of specialist circles. The common names in this book are for the most part taken from the *Checklist of the Vascular Plants of Wisconsin* (Wetter et al. 2001). Most will frankly not be of great use in talking about the plants—most common names for sedges vary greatly from region to region and are not applied consistently even within regions—but they are provided because they may be helpful in learning the plants.

Botanists and naturalists rely primarily on scientific (Latin) names for communicating about plants. A full scientific name for a species has three parts: the genus name, the specific epithet, and the authority. The genus name for the sedges in this book is *Carex*. The specific epithet for the well-known Pennsylvania sedge is *pensylvanica*. The authority is Lamarck, who published the formal scientific description of the species. The species name, then, is *Carex pensylvanica* Lamarck. (Note that the name is not "*Carex pennsylvanica*" with two *n*'s; Lamarck was French, and he consequently would have referred to Pennsylvania as "la Pensylvanie.") The name can also be abbreviated as *C. pensylvanica* Lamarck if the genus is unambiguous in context, and the authority can be abbreviated as "Lam." following a standardized reference to plant authors. One frequently does not include the authority in discussing a species, but for the sake of precision it should be included the first time a species name is used in a given context. The reason for this is that many Latin binomials have been applied (inappropriately) by more than one author in describing different species. Using authority names, however, can become unwieldy. In this book authorities are used only in part 1 and when additional names are introduced in part 2. Scientific names follow the nomenclatural treatment in *FNA*.

Usage of plant names changes as taxonomists' understanding of plants changes. Such changes typically occur for one of three reasons. First, our understanding of species boundaries often changes as we gather more data or study existing collections more closely (see discussion below). Thus, plants that were once believed to comprise separate species may with additional study all be considered part of a single species, and plants that previously were recognized as a single species may be separated into two or more. Second, taxonomists often disagree on the taxonomic rank at which a group of plants should be recognized. This book is concerned with only a few ranks: genus, species, subspecies, and variety. It is not uncommon for plants to initially be described as a new taxon at the rank of subspecies or variety and then subsequently to be elevated to species rank with additional work.

This reflects in part taxonomists' increasing understanding of the organisms they work with. Finally, names change when taxonomists find errors. A species may be described in publication more than once, for instance, by two separate authors; in such cases, the earlier name takes precedence, as long as it is valid according to the International Code of Botanical Nomenclature.

Any time a species has been recognized under more than one name, the names are referred to as synonyms of one another. Common synonyms are presented in brackets in parts 1 and 2 of this book. The *Checklist of the Vascular Plants of Wisconsin* (Wetter et al. 2001) contains a very thorough list of synonyms.

SPECIES CIRCUMSCRIPTION

This book follows the decisions published in *FNA* about what constitutes a legitimate species, subspecies, or variety. The question of how one should decide what constitutes a "good" species often arises in taxonomically difficult groups of organisms. The answer is not straightforward. The treatment in *FNA* represents a snapshot of our current understanding of North American sedge taxonomy, directly reflecting the work of approximately twenty practicing taxonomists and scores of additional researchers who have left their marks (and names) on sedge taxonomy over hundreds of years.

Although every researcher favors a slightly different mix of evidence in favor of recognizing two sets of organisms as separate species, it is probably fair to say that nearly all of the species recognized in this book are distinct in morphological characteristics, habitat, and/or geographic distribution. It is also fair to say that although very few have been studied using modern biosystematic methods that are used to assess patterns of gene flow, interspecies hybrids are the exception rather than the rule in the genus. Twenty years hence, taxonomists will almost certainly have come to the conclusion that some of the species recognized in this book ought to be recognized at a different rank (e.g., some species will come to be viewed as subspecies, some varieties will be viewed as species, etc.), while other species should be teased apart. For the time being, however, the taxonomy in this book represents a reasonable account of sedge taxonomy as it is generally accepted today.

CLASSIFICATION

Since Darwin, the understanding that all living things on Earth are descended from a common ancestor has been supported by a mass of scientific data, so that almost no one who has looked carefully at the evidence doubts that all of Earth's organisms are interrelated. This set of relationships, colloquially referred to as the Tree of Life, has become illuminated with increasing clarity in the past twenty years with the availability of molecular genetic data. Relationships throughout much of the tree are known very well. These relationships are expressed with varying degrees of adequacy in the traditional classification system handed down to us in modified form from Linnaeus.

Sedges are very closely related to rushes (Juncaceae) and somewhat more distantly related to grasses (Poaceae), bromeliads (Bromeliaceae), and a number of other monocotyledonous families. Within a modified Linnaean classification system sedges are placed as follows:

Domain: Eukaryota—the eukaryotes, complex creatures composed of cells that live in symbiosis (tight, cooperative relationship), with compartmentalization of cellular functions into different organelles; ca. 1.5 million described species
  Kingdom: Plantae—green algae and the land plants; ca. 350,000 described species
    Division: Magnoliophyta—the flowering plants; ca. 265,000 described species
      Class: Liliopsida—the monocots, a clade of flowering plants characterized by, among other things, solitary seed leaves (cotyledons); ca. 58,000 described species
        Order: Poales—a clade of monocots that includes the grasses, sedges, rushes, cattails (Typhaceae), pipe-worts (Eriocaulaceae), and numerous other families; ca. 18,000 described species
          Family: Cyperaceae—the sedge family; ca. 5,000 described species
            Genus: *Carex* L.—the genus most commonly referred to as "the sedges"; ca. 2,000 described species
              Species: *Carex pensylvanica* Lamarck—one of the most widespread carices in eastern North America

Below the genus level, *Carex* is divided into subgenera, which are themselves divided into sections. Wisconsin's *Carex* species are easily separated into two subgenera: subgenus *Vignea*, with stigmas bifid (forked like a snake's tongue) and spikes sessile, typically uniform and bisexual; and subgenus *Carex*, with stigmas bifid or, more commonly, trifid (divided into three sections) and spikes stalked, typically unisexual or strongly dominated by flowers of a single sex, the sexes thus unevenly divided between different spikes on the same plant. Learning to recognize the two subgenera quickly turns a genus of over 150 Wisconsin species into two much more manageable groups, one of approximately 50 species and one of approximately 100.

Sectional taxonomy of the genus is in a state of flux as new morphological and molecular data refine our understanding of sectional boundaries. But most of the more prominent sections and clusters of sections can be learned readily, and understanding how to recognize them makes learning the genus much more manageable. The inflated, strongly toothed perigynia of the "bladder sedges"; the pubescent perigynia, fibrous bases, and well-developed rhizomes of the *Carex pensylvanica* group; the winged perigynia of section *Ovales;* and many other distinctive morphological features are evolutionary imprints that mark groups of closely related species. While there has been much convergence in the evolution of the genus, learning to recognize the various sections is the first step to understanding the relationships, classification, and, ultimately, evolution and ecology of sedges.

Studying Sedges in the Field

Having access to a dissecting microscope is helpful in learning and identifying sedges and will be essential for identifying some species. However, most North American sedges can be learned in the field with a hand lens and a little patience. The following recommendations should make for more enjoyable study.

GO SEDGING AT AN APPROPRIATE TIME OF YEAR

Many key characters for the genus are found on the perigynium, which swells as the achene inside ripens. For this reason, sedges are most easily identified not when they are in flower but when the fruits are ripe. Most of our *Carex* species are readily identifiable in mid-June. The two major exceptions include species of section *Acrocystis* (Pennsylvania sedge and relatives), which are identifiable in May and drop their fruits by the time most other members of the genus are becoming identifiable; and species of section *Ovales* (*Carex scoparia, C. bicknellii, C. tenera*, and their relatives), which ripen in late June and July.

Sterile plants can sometimes be identified, and this book endeavors to point out reliable vegetative characters. Immature plants can also often be identified by someone familiar with the genus, but learning sedges from immature material is very difficult. To learn the genus, begin your studies with whole plants in good condition, with ripe perigynia.

DON'T BE AFRAID TO BREAK OFF A FEW PERIGYNIA

Studying sedges requires a little patience and close observation. Bring portions of the plant to your eye rather than crouching for half an hour, shading the plant you are studying and giving yourself a backache. Removing a few perigynia from an inflorescence will not kill the plant. For that matter, do not be afraid to break off a culm or vegetative shoot if the plant you are study-ing is robust and the population healthy. Sedges are perennial plants that are adapted to grazing, and they will generally not be harmed by the removal of a shoot or two. In this way, studying sedges (and, for that matter, grasses and rushes) is unlike studying trilliums or lilies. Removing a single sedge culm rarely does the entire plant any harm, whereas plucking a trillium or lily may deprive the plant of resources it needs to fruit in the following year.

USE A HAND LENS

Many of the characteristics referred to throughout this guide can be learned without a microscope, but they are easiest to see with a little magnification. If you do not own a 10× hand lens (a loupe), purchase or borrow one. These can be acquired for less than ten dollars and will make your sedge studies much more profitable and enjoyable. Bring the hand lens right to your eye, then bring the plant you are studying to the hand lens. The hand lens will serve you best if you think of it less as a magnifying glass (used at a com-fortable reading distance) than as an extension of your eye (or, in my case and perhaps yours, as a second pair of glasses). With practice, you can iden-tify many of Wisconsin's sedges from a distance, but learning the species re-quires close inspection.

CARRY A SMALL PLASTIC OR METAL RULER WITH
ENGLISH AND METRIC RULINGS

Many key characters for distinguishing closely related species are measure-ments that are not easily estimated.

IF YOU ARE COLLECTING, COLLECT ADEQUATELY

It will often be advantageous to collect material in the field for inspection at home or even sitting comfortably at the edge of the trail. *If you choose to collect sedges, be sure to collect only from populations that are unprotected (e.g., not in state natural areas or a nearby arboretum) and represented by a good number of individual plants. If you have doubts about whether you may collect in a given area, ask the land manager or owner.* For learn-ing purposes, it will generally suffice to collect one or two culms. Plants can be placed in an airtight plastic bag for short-term study or a plant press if you want to hang onto them for longer than a day or two. Make sure to remove the culm down as close to the ground as you can. The base of the plant is generally helpful and often crucial in making correct identifications. Make sure as well to note underground structures. Does the plant have long rhizomes? Take note of general growth form and habitat. Does the plant grow clonally? Are clumps large or small, or do plants occur singly? What kind of forest are you in? What plants are growing with your sedge? The more information you write in your notebook, the more clues you will have when you later sit down to study your plant.

USE MORE THAN ONE SEDGE REFERENCE, AND VISIT AN HERBARIUM IF YOU HAVE ACCESS TO ONE

No single book encompasses the experience of the many, many excellent botanists who love and have written about sedges. Go to a library and check out some of the references listed at the back of this book. Better yet, the herbarium—a museum of pressed plants and the haven of the field botanist— at your local university or college may have room for plant enthusiasts to work. Comparing your identifications against others' is invaluable at all stages of learning and research.

PART 1

Keys and Abbreviated Descriptions of
Wisconsin Carex Sections and Species

Key to Subgenera
of *Carex*

Learning to distinguish the major *Carex* subgenera—subgenus *Carex* and subgenus *Vignea*—makes learning sedges much easier. The subgenera are easy to recognize in the field once you are familiar with them, but numerous characters are needed to distinguish the two in a dichotomous key. You will save yourself time and frustration if you study this key to subgenera for five or ten minutes before you start working on any plants. It may be helpful to look through the illustrations for *Carex* subgenus *Carex* [species 1–99] and subgenus *Vignea* [100–157] to get a feel for how the perigynia, spikes, and entire inflorescence of the two subgenera differ. Doing so will place the terminology into the context of the whole plant.

Spike solitary
    Stigmas 3; achenes round or triangular in cross-section . . . . Subgenus *Carex*
        (in part), Key A
      Stigmas 2; achenes lenticular (biconvex) (subgenus *Vignea* in part)
        Rhizomes long, conspicuous, often creeping; shoots typically arising
        singly or few together; leaf blades threadlike, (less than or equal to)
        1 mm wide; spike typically unisexual, occasionally androgynous;
        perigynia (on pistillate plants) biconvex, margins rounded, beak
        0.5 mm long; calcareous wetlands, primarily northeastern quarter
        of Wisconsin (section *Physoglochin*) . . . . 101. *C. gynocrates*
        Rhizomes short and/or inconspicuous; plants cespitose; leaf blades
        involute, (less than or equal to) 1.5 mm wide; spike typically
        bisexual, occasionally unisexual; perigynia (on pistillate plants)
        planoconvex, margins acute, beak 0.5–1.5 mm long; bogs of the
        Apostle Islands and Door County (section *Stellulatae* in part) . . . . 131. *C. exilis*
        (in part)
Spikes more than 1
    Inflorescence bisexual; spikes on a given plant obviously of two kinds—
        some all or predominantly staminate, some all or predominantly pistillate—
        typically elongate, stalked; stigmas 3 per flower and achenes round or
        triangular in cross-section, or stigmas 2 and achenes lenticular but
        spikes elongate, stalked, and divided by sex as described above;
        bracts often foliose and/or sheathing; basal leaf sheaths often red
        or purple . . . . Subgenus *Carex*—Keys B–E

Inflorescence bisexual or unisexual; spikes on a given plant typically all similar, sessile, mostly bisexual (inflorescence unisexual in *Carex praegracilis* [100]; a mix of staminate, pistillate, and/or bisexual spikes typical in the inflorescence of *C. sartwellii* [102] and *C. siccata* [103]); stigmas 2 and achenes lenticular; bracts rarely foliose, not sheathing; basal leaf sheaths rarely reddish, though they may be darkened .... Subgenus *Vignea*—Keys F–J

Key to *Carex* Subgenus *Carex*

I. Spike solitary, terminal, bisexual, the staminate portion typically less
conspicuous than the pistillate portion; lateral spikes sometimes present
near the base of the plant . . . . Key A
I. Spikes 2 or more, terminal and lateral, staminate, gynecandrous, or
androgynous; at least some lateral spikes borne on the upper half of the culm
  II. Achene apex abruptly narrowed to the style base, easily distinguished
    from it; style base narrow, fragile, sometimes circled with a ring of
    constricted, inflated, or colored tissue, the texture or color typically
    setting the style base apart from the achene apex; style soon withering,
    deciduous; perigynia various, rarely inflated or borne in bottlebrush-
    like pistillate spikes; perigynium beak teeth soft, typically < 0.2 mm long
      III. Stigmas 3; perigynia terete to triangular in cross-section, occasionally
        biconvex or planoconvex and distinctly beaked; achenes triangular
        in cross-section
        IV. Perigynia pubescent . . . . Key B
        IV. Perigynia glabrous, though often the margins and occasionally the
          veins are scabrous . . . . Key C
      III. Stigmas 2; perigynia biconvex or globose, beakless to short-beaked;
        achenes lenticular, biconvex . . . . Key D
  II. Achene apex tapering gradually to the style base so that it is difficult to
    pinpoint where the style and achene meet; style base tapering, firm,
    similar in texture and color to the achene apex; style persistent;
    perigynia 2.5–20 mm long, inflated and/or pistillate spikes resembling
    bottlebrushes; perigynium beak teeth firm, > 0.2 mm long . . . . Key E

KEY A—UNISPICATE "EUCARICES"

Plants in this group have a solitary bisexual spike at the tip of each culm.
Plants in section *Phyllostachyae* often have additional lateral spikes very low
on some culms, but these are usually inconspicuous. Inflorescences of our
unispicate species typically have few perigynia, and the staminate portions
of the spike are relatively inconspicuous. All are androgynous except *Carex
typhina* [74], of which some culms produce solitary, terminal gynecandrous
spikes.

1. Perigynia pubescent at least sparsely
  2. Perigynia ≤ 3.2 mm long, beak ≤ 1 mm long; dry calcareous prairies
  and limestone outcrops, occasional in dry sands and wet calcareous
  prairies . . . . 10. *C. umbellata* (in part)
  2. Perigynia ≥ 3.1 mm long, beak ≥ 0.9 mm long; dry sandy soils
    3. Perigynium bodies sparsely pubescent near the beak, otherwise smooth;
    leaf blades pale green, tough, short . . . . 11a. *C. tonsa* var. *tonsa* (in part)
    3. Perigynium bodies pubescent throughout; leaf blades bright green,
    soft, longer than culms . . . . 11b. *C. tonsa* var. *rugosperma* (in part)
1. Perigynia glabrous
  4. Lowest pistillate scale of the terminal spike foliose, exceeding the spike;
  lateral spikes, if present, only near the base of the plant (section
  Phyllostachyae)
    5. Bractlike pistillate scales < 3 mm wide with spreading margins, not
    concealing the perigynia . . . . 1. *C. jamesii*
    5. Bractlike pistillate scales > 3 mm wide, partly wrapped around the
    perigynia and nearly concealing them . . . . 2. *C. backii*
  4. Lowest pistillate scales shorter than the perigynia, never foliose; lateral
  spikes lacking
    6. Spike gynecandrous with a prominent staminate base, ovoid to
    plumply cylindrical, densely packed with perigynia; perigynia abruptly
    beaked; floodplain forest species along larger rivers, primarily
    southwestern Wisconsin . . . . 74. *C. typhina* (in part)
    6. Spike androgynous with a small and often inconspicuous staminate
    portion, slender, few-flowered; perigynia blunt and beakless or
    long-tapered to the apex, not abruptly beaked; peaty wetlands
      7. Perigynia reflexed at maturity, slender, long-pointed, 6–8 mm
      long; sphagnum bogs northward (section *Leucoglochin*
      [= *Orthocerates*]) . . . . 3. *C. pauciflora*
      7. Perigynia ascending at maturity, squat, blunt at the apex,
      2.5–5 mm long; fens and bogs throughout the state (section
      Leptocephalae [= *Polytrichoideae*]) . . . . 4. *C. leptalea*

Key B—"Eucarices" with perigynia pubescent, not inflated, the styles deciduous and abruptly jointed at the achene apex (i.e., not in the "bottlebrush" and "bladder" sedges group)

Perigynium pubescence has arisen independently in many *Carex* lineages. Consequently, the species in this key are a morphologically heterogeneous assemblage. Among the "peachfuzz wood sedges" in our flora, only sections *Acrocystis* (e.g., Pennsylvania sedge) and *Clandestinae* (e.g., Richardson's sedge) are morphologically similar to one another. The remaining sections are quite distinct: section *Anomalae*, represented in our flora by a single species with scabrous foliage and scabrous or short-pubescent perigynia; members of wetland section *Paludosae* that have relatively narrow foliage; *Carex assiniboinensis* [43], typically classified in the *Hymenochlaenae—Sylvaticae* group but distinguished by its oblique perigynium orifice and spider plant–like style of vegetative reproduction; *C. hirtifolia* [5], which is softly pubescent on all surfaces; and members of section *Porocystis* that have pubescent perigynia.

1. Leaf blades strongly scabrous on the upper surface; basal leaf sheaths not tinged red-purple; perigynia scabrous or sparsely pubescent, hairs sometimes limited to the veins, 2-ribbed with several distinct veins, beak > 1 mm long, curved away from the spike axis; plant mostly of wet habitats in far northern Wisconsin (section *Anomalae*) . . . . . . 25. *C. scabrata* (in part— more likely to key to Key C)
1. Leaf blades not strongly scabrous on the upper surface; basal leaf sheaths often tinged red-purple; perigynia distinctly pubescent, venation subtle or concealed by the pubescence (venation visible in *Carex houghtoniana* [87]), beak < 1 mm long in most species (> 2 mm long in *C. assiniboinensis* [43]), absent in some, straight to slightly bent; habitats and distribution various
   2. Basal leaf sheaths red, ladder-fibrillose; plants loosely cespitose or shoots arising singly; rhizomes slender, elongate (section *Paludosae* in part)
      3. Perigynia ≥ 4.5 mm long, sparsely pubescent, venation and surface texture visible through the hairs; sandy and rocky soils, predominantly in the northern third of the state . . . . 87. *C. houghtoniana*
      3. Perigynia ≤ 4.5 mm long, densely pubescent, venation and other surface features obscured; wetlands throughout the state
         4. Leaf blades flat or M-shaped in cross-section except at the base and the tip, mostly > 2 mm wide . . . . 88. *C. pellita*
         4. Leaf blades involute or V-shaped in cross-section, elongated to a wiry tip, mostly ≤ 2 mm wide . . . . 89. *C. lasiocarpa*
   2. Basal leaf sheaths reddish, fibrous in some species (especially section *Acrocystis*, second lead of couplet 8 below) but not ladder-fibrillose; plants cespitose or culms arising singly; rhizomes various
      5. Leaf blades, sheaths, and culms pubescent
         6. Perigynium beaks ≥ 1 mm long; terminal spike staminate; leaf blades 2.5–7 mm wide; locally abundant in rich deciduous forests throughout most of the state (section *Hirtifoliae*) . . . . 5. *C. hirtifolia*
         6. Perigynia beakless; terminal spike gynecandrous; leaf blades 1.5–3 mm wide; restricted to a few sites in southeastern and central Wisconsin (section *Porocystis* in part) . . . . 21. *C. swanii*
      5. Leaf blades, sheaths, and culms glabrous
         7. Arching vegetative shoots (stolons) elongating to 1 m or longer by late summer, bearing reflexed leaf blades, producing plantlets at the tips in mid- to late summer in the manner of a spider plant; perigynia tapering gradually to a straight, elongate beak, the orifice cut obliquely (at an angle), forming a slit down the side of the beak; pistillate spikes linear, sometimes arching or drooping; perigynia separated, typically only the beaks overlapping the perigynium bases immediately above . . . . 43. *C. assiniboinensis* (in part)
         7. Specialized vegetative shoots lacking; perigynia short-beaked, the orifice cut ± perpendicular to the axis of the beak; pistillate

spikes ovoid to oblong, ascending or the lateral spikes pendulous
on long, slender stalks; perigynium bodies overlapping

8. Perigynium beaks generally ≤ 0.5 mm long, untoothed (section *Clandestinae*
[= *Digitatae*])

  9. Older leaf tips brown, separated from the green blades by a red-purple
line; lowest pistillate spike long-stalked, arising from a lower node;
terminal spike typically androgynous; common in northern forests,
occasional in southern mesic forests . . . . 6. *C. pedunculata*

  9. Older leaf tips green or brown, not separated from the green blades
by a red-purple line; lowest pistillate spike sessile or short-stalked,
arising from an upper node; terminal spike staminate; mostly open
calcareous habitats, distribution various

    10. Plants densely cespitose, short-rhizomatous; staminate spikes
3–7 mm; dolomitic wetlands and boreal forests of Door County and
the Apostle Islands . . . . 7. *C. concinna*

    10. Plants loosely cespitose, long-rhizomatous; staminate spikes
10–25 mm; calcareous prairies, occasional in barrens and calcareous
wetlands, primarily in the southern third of the state and Brown
County . . . . 8. *C. richardsonii*

8. Perigynium beaks > 0.5 mm long, bidentate (section *Acrocystis* [= *Montanae*])

  11. Pistillate spikes borne both near the base of the plant on slender,
short to elongate stalks, sometimes partly concealed by the leaf bases,
and near the tips of the longer culms

    12. Basal leaf sheaths not or only slightly fibrous; pistillate scales
shorter than perigynia; lowest bract on elongated culms equaling
or exceeding the tip of the terminal spike; uncommon in northern
and central Wisconsin . . . . 9. *C. deflexa* (in part)

    12. Basal leaf sheaths fibrous; pistillate scales nearly equaling to
exceeding perigynia; lowest bract on elongated culms shorter than
the inflorescence; distributions various

      13. Perigynia 2.2–3.2 mm long, pubescent overall, beak 0.4–1 mm
with teeth 0.1–0.2 mm long; calcareous prairies and limestone
outcrops, occasional in dry sands . . . . 10. *C. umbellata* [= *C. abdita*]

      13. Perigynia 3.1–4.7 mm long, pubescent overall [11a] or nearly
glabrous and pubescent only at the base of the beak [11b], beak
0.9–2 mm with teeth 0.2–0.5 mm long; dry sandy soils

        14. Perigynium bodies sparsely pubescent at the base of the
beak, otherwise glabrous; leaf blades pale green, short,
tough, strongly scabrous . . . . 11a. *C. tonsa* var. *tonsa*

        14. Perigynium bodies pubescent overall; leaf blades bright
green, longer than culms, relatively soft . . . . . 11b. *C. tonsa* var. *rugosperma*
[= *C. rugosperma*]

  11. Pistillate spikes borne near the tips of the culms, close to the staminate
spikes, none near the base of the plant

    15. Plants densely cespitose; rhizomes thick, short; widest leaf blades
≥ 3 mm wide; basal leaf sheaths little if at all fibrous; common
in northern forests . . . . 12. *C. communis*

    15. Plants loosely cespitose (densely cespitose in *C. albicans* [18]);
rhizomes typically slender, elongate, creeping; leaf blades mostly
< 3 mm wide; basal leaf sheaths fibrous in most species; habitats
and distributions various

16. Perigynium bodies approximately globose
    17. Perigynium beaks ≥ 0.9 mm long, ≥ half as long as the body;
    culms usually strongly scabrous beneath the inflorescence;
    rare in northeastern Wisconsin . . . . 13. *C. lucorum*
    17. Perigynium beaks ≤ 0.9 mm long, < half as long as the body;
    culms smooth or weakly scabrous beneath the inflorescence;
    common throughout the state
        18. Perigynia mostly ≥ 1.5 mm wide; achenes mostly >
        2 mm long . . . . 14. *C. inops*
        18. Perigynia mostly ≤ 1.5 mm wide; achenes mostly
        < 2 mm long . . . . 15. *C. pensylvanica*
16. Perigynium bodies ellipsoid
    19. Perigynia longer than the pistillate scales, conspicuous
        20. Culms slender, the longer ones frequently arching or
        curved; leaf blades mostly equaling or exceeding the
        culms; bract subtending the lowest spike equaling or
        exceeding the tip of the terminal spike; perigynia
        ≤ 3.1 mm long . . . . 9. *C. deflexa* (in part)
        20. Culms straight, erect; leaf blades shorter than the culms;
        bract subtending the lowest spike shorter than the
        inflorescence; perigynia ≥ 3.2 mm long . . . . 16. *C. peckii*
    19. Perigynia shorter than the pistillate scales, largely
    concealed by them
        21. Lowest 2 pistillate spikes usually separated by ≥ 7 mm;
        bract subtending the lowest spike equaling or exceeding
        the tip of the terminal spike; known from two sites in
        far northern Wisconsin . . . . 17. *C. novae-angliae*
        21. Lowest 2 pistillate spikes close to one another or
        overlapping; bract subtending the lowest spike shorter
        than the inflorescence; uncommon in central
        Wisconsin
            22. Staminate scales near the middle of the terminal
            spike usually bristle-tipped, the midrib extending to
            the tip and often protruding beyond; culms typically
            arching or spreading; infrequent, central
            Wisconsin . . . . 18a. *C. albicans* var. *emmonsii*
                                         [= *C. emmonsii*]
            22. Staminate scales near the middle of the terminal
            spike acute or blunt at the apex, not bristle-tipped,
            the midrib not extending to the tip; culms typically
            erect or ascending; known only from Devil's Lake
            State Park . . . . 18b. *C. albicans* var.
                                      albicans [= *C. artitecta*]

KEY C—MISCELLANEOUS "EUCARICES" WITH UNINFLATED, GLABROUS PERIGYNIA

1. Lowest pistillate scales exceeding the tip of the terminal spike, foliose;
terminal spike androgynous; lateral spikes, if present, at the plant base
(section *Phyllostachyae*)

2. Bractlike pistillate scales 1.5–3 mm wide; scale margins spreading,
    not concealing the perigynia . . . . 1. *C. jamesii*
2. Bractlike pistillate scales > 2.5 mm wide; scale margins partly wrapped
    around the perigynia, concealing them or nearly so . . . . 2. *C. backii*
1. Lowest pistillate scales shorter than or slightly exceeding the perigynia,
    not approaching the tip of the spike, scalelike; terminal spike generally
    staminate; lateral spikes on the upper half of the culm
    3. Bract subtending the lowest spike sheathless or with sheath < 1 mm long
        4. Roots covered with a yellow, feltlike tomentum (may be faded in
            herbarium material); lateral spikes pendulous on slender stalks;
            pistillate scales as long as perigynia or longer; bogs (section *Limosae*)
            5. Pistillate scales wider than perigynia; leaf blades 1–2.5 mm wide,
                margins involute; culms mostly aphyllopodic . . . . 19. *C. limosa*
            5. Pistillate scales narrower than perigynia; leaf blades 1–4 mm wide,
                margins revolute; culms mostly phyllopodic . . . . 20. *C. magellanica*
        4. Roots without yellow tomentum; lateral spikes erect to ascending,
            occasionally drooping (*Carex scabrata* [25]); pistillate scales shorter
            than to nearly equaling the perigynia in length; habitats various
            6. Leaf blades and/or sheaths pubescent (section *Porocystis* in part)
                7. Terminal spike gynecandrous, > three-tenths of flowers pistillate;
                    known from a single collection in Iowa County . . . . 22. *C. bushii*
                7. Terminal spike usually staminate, less often gynecandrous,
                    ≤ one-quarter of flowers pistillate
                    8. Perigynia 1–1.5 mm wide, beakless; locally common in far
                        north-central Wisconsin and two localities in southeastern
                        Wisconsin . . . . 23. *C. pallescens*
                    8. Perigynia 1.5–2.2 mm wide, short-beaked; uncommon in
                        southeastern and western Wisconsin . . . . 24. *C. torreyi*
            6. Leaf blades and sheaths glabrous (upper surface of the blades
                scabrous in *Carex scabrata* [25])
                9. Perigynia scabrous or sparsely pubescent/bristly on at least
                    some of the veins, distinctly 2-ribbed, beaks > 1 mm long; leaf
                    blades scabrous on the upper surface; perigynia olive green,
                    not contrasting with the pistillate scales; terminal spike staminate,
                    rarely gynecandrous (section *Anomalae*) . . . . 25. *C. scabrata* (in part)
                9. Perigynia smooth, veinless or several-veined, beaks 0–1 mm
                    long (to 2 mm long in *Carex vaginata* [29]); leaf blades smooth
                    or barely scabrous on the upper surface; perigynia pale gray
                    green to blue-green, contrasting with the dark pistillate scales;
                    terminal spike gynecandrous, rarely staminate (section
                    Racemosae [= *Atratae*])
                    10. Plants long-rhizomatous, arising one to few in a clump;
                        basal leaf sheaths ladder-fibrillose (forming a distinctive,
                        ladderlike reticulum of horizontal fibers on the inner band
                        as the sheath splits open); spikes 2–5, 10–25 mm long,
                        6–10 mm wide; perigynia broadest below the middle;
                        calcareous wetlands throughout much of the state . . . . 26. *C. buxbaumii*
                    10. Plants short-rhizomatous, cespitose; basal leaf sheaths not
                        fibrous or ladder-fibrillose; spikes usually 3, 5–12 mm long,
                        3–6 mm wide; perigynia broadest near the middle; algific
                        talus slopes in Grant County . . . . 27. *C. media*

3. Lowest bract forming a closed, tubular sheath > 1 mm long (bladeless in *Carex eburnea* [28] and *C. plantaginea* [59])

11. Plants long-rhizomatous or stoloniferous, shoots typically arising singly or few to a clump (cespitose and forming sods in *Carex eburnea* [28])

12. Leaf blades involute, wiry, < 1 mm wide; plants forming dense sods; calcareous soils, most common in juniper glades (section *Albae*) . . . . 28. *C. eburnea*

12. Leaf blades flat, V-shaped or W-shaped in cross-section, > 1 mm wide; plants colonial but not typically forming dense mats; mesic to wet calcareous habitats (except *Carex meadii* [32], which favors dry lime prairies)

13. Perigynia and pistillate scales flecked with red (10× magnification), packed densely into cylindrical pistillate spikes like corn kernels on a cob; calcareous wetlands, primarily in Door County and northeastern Wisconsin (section *Granulares* in part) . . . . 53. *C. crawei*

13. Perigynia and pistillate scales not red-flecked or packed densely into cylindrical pistillate spikes; habitats and distributions various (section *Paniceae*)

14. Perigynium beak 0.5–2 mm long, straight; white cedar swamps, northeastern Wisconsin and Douglas County . . . 29. *C. vaginata*

14. Perigynium beak 0–0.5 mm long, bent in most species (absent or straight in *Carex livida* [31])

15. Lower leaf sheaths bladeless, a beautiful burgundy; wet to mesic forests . . . . 30. *C. woodii*

15. Lower leaf sheaths mostly bearing blades, not strongly colored; open prairies, fens, and bogs

16. Perigynium beak straight or absent; leaf blades leathery, involute; northern fens, primarily along Lake Superior . . . . 31. *C. livida*

16. Perigynium beak very short, bent; leaf blades herbaceous, flat; southern calcareous fens and prairies

17. Leaf blades gray, stiff; ligules ≤ 1.2 times as long as wide; achenes ≥ 1.7 mm wide; most common in dry calcareous prairies, occasional in fens and calcareous wet prairies . . . . 32. *C. meadii*

17. Leaf blades green, relatively soft; ligules 1–2 (occasionally 0.8) times as long as wide; achenes ≤ 1.6 mm (occasionally 1.8 mm) wide; most common in fens and wet prairies . . . . 33. *C. tetanica*

11. Plants short- or inconspicuously rhizomatous, cespitose

18. Perigynia spreading or, more commonly, reflexed, beaks elongate, bidentate, straight or bent away from the spike axis; wet calcareous mineral soils (section *Ceratocystis*)

19. Pistillate spikes 4–7 mm thick; perigynia ≤ 4 mm long, beak much shorter than body, straight or only slightly bent away from the spike axis . . . . 34. *C. viridula*

19. Pistillate spikes 7–12 mm thick; perigynia ≥ 3.5 mm long, beak ≥ half as long as body, bent away from the spike axis at least 15° to the body

20. Perigynium beaks smooth to the tip; fresh pistillate
scales yellowish green, the same color as the perigynia
and soon hidden by them . . . . 35. *C. cryptolepis*
20. Perigynium beaks conspicuously scabrous at least near
the tip (20× magnification); fresh pistillate scales brownish,
contrasting with the perigynia and conspicuous until the
lower perigynia reflex . . . . 36. *C. flava*
18. Perigynia ascending to occasionally spreading, beak length various,
straight or bent; habitats various, most common in woodlands
and forests
21. Pistillate spikes mostly pendulous on narrow, elongate stalks, narrowly
oblong to slenderly cylindrical, generally elevated well above the leaf tops
on tall culms (section *Hymenochlaenae*)
22. Perigynia tapered to a curved or bent beak, 2-ribbed but otherwise
veinless; bracts sheathless; shaded seeps of the Apostle Islands, Baraboo
Hills and Blue Hills, and Rock Island (Door County) . . . . 37. *C. prasina*
22. Perigynia straight-beaked or beakless, mostly several-veined (2-veined
in *Carex sprengelii* [41]); bracts long-sheathing
23. Terminal spike gynecandrous (section *Hymenochlaenae—
Gracillimae* group)
24. Perigynia beakless; leaf blades and sheaths glabrous . . . . 38. *C. gracillima*
24. Perigynia tapered to a distinct beak; leaf blades and/or
sheaths pubescent
25. Pistillate scales long-awned, exceeding the perigynium tips;
floodplains, southwestern Wisconsin . . . . 39. *C. davisii*
25. Pistillate scales at most short-awned, shorter than the
perigynia; threatened species of mesic to wet forests,
sporadic in the eastern quarter of the state . . . . 40. *C. formosa*
23. Terminal spike staminate
26. Basal leaf sheaths drab or brown, fibrous, bearing green blades;
pistillate spikes ovate to cylindrical
27. Perigynium body globose, beak cylindrical, approximately
as long as the body; basal leaf sheaths strongly fibrous;
leaf blades 2–4 (mostly ≥ 2.5) mm wide; culms to 1 m long;
widespread (section *Hymenochlaenae—Longirostres*
group) . . . . 41. *C. sprengelii*
27. Perigynium body lance-ovoid, beak tapering, half as long
as the perigynium body; basal leaf sheaths weakly fibrous;
leaf blades 1–4 (mostly ≤ 2) mm wide; culms to 0.5 m long;
Door and Bayfield counties (section *Chlorostachyae*
[= *Capillares*]) . . . . 42. *C. capillaris*
26. Basal leaf sheaths typically purplish, bladeless or short-bladed
(blades on the lowest sheaths generally < 2 cm), not fibrous;
pistillate spikes mostly slender, elongate (section
Hymenochlaenae—Sylvaticae group)
28. Vegetative shoots elongating to ≥ 1 m, bearing reflexed
leaf blades, producing plantlets at the tips in mid- to
late summer in the manner of a spider plant; perigynia
tapering gradually to a straight beak 2.5 mm long,
perigynium beaks overlapping bodies of the perigynia
above them . . . . 43. *C. assiniboinensis* (in part)

28. Vegetative shoots not specialized as above; perigynia
    tapering to a beak 0.2–2 mm long; perigynium bodies
    overlapping
    29. Leaf blades and sheaths pubescent; pistillate spikes
        narrowly oblong, 4–5 mm thick; northeastern and
        far northern Wisconsin . . . . 44. *C. castanea*
    29. Leaf blades and sheaths glabrous; pistillate spikes
        cylindrical, 2–4 mm thick; widespread in central and
        northern Wisconsin
        30. Perigynia ≤ 5 mm long, short-beaked, abruptly
            constricted at the base to a short stipe; common
            in northern forests . . . . 45. *C. arctata*
        30. Perigynia ≥ 5 mm long, long-beaked, tapering to a
            slender, acute base; wet forests and woodlands,
            central and northern Wisconsin . . . . 46. *C. debilis*
21. Pistillate spikes ascending or the lowest pendulous, broadly oblong
    to short-cylindrical (may be cylindrical in *Carex laxiflora* [52]),
    elevated above the leaf tops on culms that exceed the leaf tops not at
    all to greatly
    31. Basal leaf sheaths bladeless (culms aphyllopodic), red-purple;
        perigynia sparsely pubescent, occasionally glabrous; common in
        northern forests, occasional in southern forests (section
        Clandestinae in part) . . . . 6. *C. pedunculata* (in part)
    31. Basal leaf sheaths bearing blades (culms phyllopodic), pale, green,
        brown, or red- to purple-tinged; perigynia glabrous; distributions
        various
        32. Culms acutely triangular in cross-section, weak and easily
            compressed, typically flattened in pressed material, the angles
            narrowly winged in some species (section *Laxiflorae*)
            33. Widest leaf blades mostly > 1 cm wide; bracts wide; concealing
                the spikes; pistillate scale apex obtuse or truncate, occasionally
                with a short, abrupt, apical tooth . . . . 47. *C. albursina*
            33. Widest leaf blades < 1 cm wide; bracts not concealing the
                spikes; pistillate scale apex acute, toothed or short-awned
                in some species
                34. Staminate spike long-stalked, raised well above the pistillate
                    spikes; basal leaf sheaths subtly purple-tinged . . . . 48. *C. gracilescens*
                34. Staminate spike short-stalked, its base generally buried
                    among the uppermost pistillate spikes (occasionally stalked
                    in *Carex blanda* [50]); basal leaf sheaths pale to brownish
                    (obscurely purple-tinged in *C. ormostachya* [51])
                    35. Perigynium veins 8–18, mostly faint; common in
                        northern Wisconsin . . . . 49. *C. leptonervia*
                    35. Perigynium veins 22–32, distinct; abundance and
                        distributions various
                        36. Margins of the uppermost bract sheath finely
                            scabrous; common in southern Wisconsin . . . . 50. *C. blanda*
                        36. Margins of the uppermost bract sheath smooth;
                            local in relatively few counties
                            37. Perigynia ≤ 3.3 mm long, beak < 0.5 mm long;
                                pistillate scales 2–2.5 mm wide, apex abruptly

      sharp- or bristle-tipped; fresh basal leaf sheaths
      brown to purplish; Bayfield and Door counties,
      northeastern Wisconsin . . . . 51. *C. ormostachya*
    37. Perigynia ≥ 3.2 mm long, beak ≥ 0.5 mm long;
      pistillate scales 1–1.5 mm wide, apex obtuse
      to acute; fresh basal leaf sheaths brown to pale;
      beech forests along Lake Michigan . . . . 52. *C. laxiflora*
32. Culms obtusely triangular in cross-section, firm
    38. Perigynia > 25 per lateral spike; perigynium veins raised in
      both fresh and dried material; perigynium surface and
      pistillate scales flecked with yellow to red dots and short
      dashes (visible at 10× in bright light); open to wooded
      calcareous wetlands (section *Granulares* in part) . . . . 54. *C. granularis*
    38. Perigynia < 25 per lateral spike; perigynium veins impressed
      at least in fresh material; perigynium surface and pistillate
      scales not flecked with pigment; habitats various
      39. Perigynia terete or obtusely triangular in cross-section;
        nerves impressed in both fresh and dry perigynia; wet to
        mesic woods (section *Griseae*)
        40. Perigynia loosely enveloping the achene, terete or angles
          rounded, beakless or scarcely beaked
          41. Staminate spike base elevated above the pistillate
            spikes; stalks of lateral spikes finely scabrous;
            perigynia ≤ 4 mm long; wet prairies, fens, and
            degraded calcareous wet meadows . . . . 55. *C. conoidea*
          41. Staminate spike generally overlapped by the upper
            pistillate spikes; stalks of lateral spikes smooth
            or finely scabrous; perigynia 4.5–5.5 mm long;
            wet woods . . . . 56. *C. grisea*
        40. Perigynia tightly enveloping the achene, obtusely
          triangular in cross-section, distinctly beaked
          42. Bract sheaths pubescent; basal leaf sheaths
            brownish . . . . 57. *C. hitchcockiana*
          42. Bract sheaths glabrous; basal leaf sheaths
            purple-red . . . . 58. *C. oligocarpa*
      39. Perigynia acutely triangular in cross-section; perigynium
        veins impressed in fresh material, raised in dry perigynia
        (section *Careyanae*)
        43. Basal leaf sheaths purple
          44. Widest leaf blades 1–3 cm wide; bract sheaths
            purple; perigynia < 5 mm long; northern half and
            west-central Wisconsin and counties adjacent to
            Lake Michigan . . . . 59. *C. plantaginea*
          44. Widest leaf blades < 2 cm (usually < 1.2 cm) wide;
            bract sheaths green or purple-striped; largest
            perigynia ≥ 5 mm long; rare in Polk, Vernon, and
            LaCrosse counties . . . . 60. *C. careyana*
        43. Basal leaf sheaths white or brown
          45. Leaf blades strongly glaucous; culms capillary
            and very weak, 0.2–0.5 mm thick; Door
            County . . . . 61. *C. platyphylla*

45. Leaf blades green or glaucous; culms 0.5–1 mm
thick; mostly southern Wisconsin, not known
from Door County
46. Lowest scale of lateral spikes empty or
subtending a staminate flower; widest leaf
blades 5–12 mm . . . . 62. *C. laxiculmis*
46. Lowest scale of lateral spikes subtending a
pistillate flower; widest leaf blades 3–5 mm . . . . 63. *C. digitalis*

KEY D—DISTIGMATIC "EUCARICES"

Individual plants with two stigmas occur sporadically in subgenus *Carex*, but only sections *Bicolores* and *Phacocystis* are consistently distigmatic. In specimens with stigmas broken off, look for the lenticular achenes that are always associated with the distigmatic condition. Learn the look of the sections by studying the illustrations of *Carex aurea* [64] (section *Bicolores*); *C. stricta* [70], *C. haydenii* [71], and *C. aquatilis* [68] (section *Phacocystis*, the old section *Acutae*); and *C. crinita* [66] (section *Phacocystis*, the old section *Cryptocarpae*).

1. Lateral spikes oblong to narrowly oblong, ascending or arching,
0.5–2 cm long; perigynia plump, obovoid; bracts typically short-sheathing;
plants mostly < 0.5 m tall (section *Bicolores*)
2. Mature perigynia orange, fleshy; lateral spikes loosely flowered,
internodes in the middle of the spike often ≥ 1 mm long; terminal spike
usually staminate, sometimes gynecandrous; calcareous wetlands,
primarily in eastern Wisconsin and northern Wisconsin near Lake
Superior . . . . 64. *C. aurea*
2. Mature perigynia whitish, dry; lateral spikes densely flowered, internodes
in the middle of the spike < 1 mm long, often < 0.5 mm long; terminal
spike gynecandrous, rarely staminate; calcareous wetlands of Door
County . . . . 65. *C. garberi*
1. Lateral spikes long-cylindrical, ascending or drooping, mostly > 2 cm long;
perigynia compressed, biconvex, lenticular; bracts sheathless; plants mostly
> 0.5 m tall (section *Phacocystis*)
3. Lateral spikes drooping and/or pendulous; pistillate scales scabrous-
awned [= *Cryptocarpae*]
4. Lower leaf sheaths smooth; apex of pistillate scale bodies notched
to truncate . . . . 66. *C. crinita*
4. Lower leaf sheaths scabrous; apex of pistillate scale bodies truncate
to acute . . . . 67. *C. gynandra*
3. Lateral spikes ascending or arching; pistillate scales unawned [= *Acutae*]
5. Lowest bract conspicuously longer than the inflorescence; ligules
longer than wide
6. Staminate spikes 1–3; pistillate spikes ≥ 4 mm wide; perigynia
veinless; common in wetlands throughout much of the state . . . . 68. *C. aquatilis*
6. Staminate spikes 1; pistillate spikes ≤ 4 mm wide; perigynia
veined; uncommon, Apostle Islands and Vilas County . . . . 69. *C. lenticularis*

5. Lowest bract shorter than to equaling the inflorescence; ligules various
  7. Lowest leaf sheaths ladder-fibrillose; ligule longer than wide
    8. Pistillate scales appressed, shorter than the perigynia; perigynia
    flat on both faces, apex tapering to beakless; forming hummocks
    in sedge meadows and marshes . . . . 70. *C. stricta*
    8. Pistillate scales spreading, exceeding the perigynium tips; perigynia
    convex on both faces, apex rounded with a very short beak; clonal
    in wet prairies, fens, sedge meadows . . . . 71. *C. haydenii*
  7. Lowest sheaths not ladder-fibrillose; ligule wider than long
    9. Pistillate scales black with a light midrib; some perigynia dark-
    spotted or mottled on the upper half; known from a few
    wetlands in Superior . . . . 72. *C. nigra*
    9. Pistillate scales pale to reddish brown; perigynia without spots;
    floodplains, mostly in the southern half of the state . . . . 73. *C. emoryi*

KEY E—"BOTTLEBRUSH" AND "BLADDER" SEDGES

This prominent group of wetland sedges is characterized by swollen, bladderlike perigynia and pistillate spikes densely packed with perigynia, resembling a bottlebrush. Some species have both characteristics; few have neither. The group is held together by a character that seems obscure at first but is worth learning: styles in all of them are, in Fernald's (1950: 300) words, "continuous, not jointed at the base, firm and persistent at summit of achene." Compare achenes of several species to understand this character. Select a common upland member of subgenus *Carex* to examine alongside a common bladder or bottlebrush sedge such as *Carex lacustris* [86], *C. hystericina* [78], or *C. intumescens* [96]. Cut open a perigynium from each plant and inspect the apex of the achene body with a hand lens or microscope (it may be difficult to detect this character in all species without a microscope or 20× hand lens and good light). In the bladder/bottlebrush sedges, the base of the style thickens gradually, tapering so that it looks like a natural and continuous outgrowth of the achene apex. The style base is the same texture and color as the achene, so it is difficult to say in bladder/bottlebrush sedges just where the achene ends and the style begins. In the other members of the genus, the style is abruptly differentiated from the achene apex. Achenes in these species typically narrow abruptly at the apex, and the style base is typically narrow and breaks off more easily than styles of the bladder/bottlebrush sedges. In some species there is a constriction or thickened ring near the style base that further distinguishes the style from the achene. There is a range of variation in this style character, but studying a handful of species will give you a feel for it.

    Even without learning the style character, you will find the major sections of bladder/bottlebrush sedges easy to recognize by their highly inflated perigynia (the "bladder" sedges) and/or divergent to reflexed perigynia packed densely into bristly, cylindrical inflorescences (the "bottlebrush" sedges). You can get a sense for the characteristics that mark these sections by studying the inflorescence and perigynium pictures in this book: *Carex comosa* [75], *C. hystericina* [78], *C. oligosperma* [80], *C. tuckermanii* [81], and *C. utriculata* [84] (section *Vesicariae*); *C. trichocarpa* [90] (section *Carex*); *C. lacustris* [86] (section *Paludosae*); and *C. grayi* [97], *C. intumescens* [96], and *C. lupulina* [98] (section *Lupulinae*). The less frequently encountered bladder sedges of sections *Rostrales* and *Squarrosae* are not illustrated.

1. Perigynium bodies widest above the middle, truncate at the top, abruptly narrow-beaked; terminal spike staminate or gynecandrous; spikes 1–several, occasionally solitary (section *Squarrosae*) . . . . 74. *C. typhina* (in part)
1. Perigynium bodies widest at or below the middle, gradually tapering to the apex, beaked or beakless; terminal spike staminate; spikes 2–several
  2. Pistillate scales with scabrous awns as long as or longer than the scale body; perigynia distinctly beaked, densely packed into bottlebrush-like pistillate spikes (section *Vesicariae* in part [= *Pseudocypereae*])
    3. Perigynium body flattened-triangular in cross-section, scarcely inflated; all or at least the lower perigynia reflexed at maturity
      4. Teeth of the perigynium beak curved outward, ≥ 1.3 mm long; spikes ≥ 12 mm thick; throughout Wisconsin . . . . 75. *C. comosa*
      4. Perigynium beak teeth straight or only slightly curved, mostly < 1.2 mm long; spikes ≤ 12 mm thick; northern half and eastern half of Wisconsin . . . . 76. *C. pseudocyperus*
    3. Perigynium body approximately terete, inflated, divergent or ascending; perigynia divergent, the lower perigynia rarely reflexed
      5. Plants long-rhizomatous, shoots arising singly; known in Wisconsin from a 1965 Iowa County collection . . . . 77. *C. schweinitzii*
      5. Plants short-rhizomatous, cespitose; more widespread
        6. Perigynium body slightly inflated, ≤ 2 mm wide, 13–21-veined; wetlands throughout the state . . . . 78. *C. hystericina*
        6. Perigynium body inflated, ≥ 2 mm wide, 7–12-veined; floodplains, mostly along larger rivers . . . . 79. *C. lurida*
  2. Pistillate scales awnless to short-awned, the awn, if present, smooth (occasionally scabrous at the tip) and shorter than the scale body; perigynia short- to long-beaked; pistillate spikes not bottlebrush-like (except *Carex retrorsa* [82], which resembles *C. hystericina* [78] but has pistillate scales with smooth margins)
    7. Perigynia ≤ 10 mm long; spikes mostly elongated (except in *Carex oligosperma* [80])
      8. Perigynium walls thin, papery; fertile shoots obvious in most populations (section *Vesicariae* in part [= *Vesicariae* in the more narrow, traditional sense])
        9. Leaf blades involute, wiry at the tips; pistillate spikes compact, sessile, globose to short-oblong, 3–15-flowered . . . . 80. *C. oligosperma*
        9. Leaf blades not involute or wiry, flat or W-shaped in cross-section; pistillate spikes distinctly longer than thick (except

Carex grayi [97]), sessile to long-stalked, oblong to narrowly
cylindrical, 15–150-flowered
    10. Achene asymmetrical, wrinkled on one side . . . . 81. *C. tuckermanii*
    10. Achene symmetrical, not wrinkled
        11. Perigynia mostly reflexed at maturity; bract of the
        lowest spike typically ≥ 3 times as long as the
        inflorescence . . . . 82. *C. retrorsa*
        11. Perigynia ascending or spreading at maturity; bract of
        the lowest spike ≤ 2.5 times as long as the inflorescence
            12. Plants densely cespitose; rhizomes short; basal leaf
            sheaths thickened but not spongy; ligules longer
            than wide . . . . 83. *C. vesicaria*
            12. Plants colonial; rhizomes long-creeping; basal leaf
            sheaths spongy-thickened; ligules as long as wide
                13. Upper (adaxial) leaf blade surfaces not papillose;
                widest leaf blades 4.5–12+ mm wide; common,
                widespread . . . . 84. *C. utriculata*
                13. Upper (adaxial) leaf blade surfaces strongly
                papillose; widest leaf blades 1.5–4.5 mm wide;
                uncommon, only in the northern third of the
                state . . . . 85. *C. rostrata*
8. Perigynium walls thick, firm; vegetative shoots often more prominent
than the fertile shoots
    14. Vegetative culms absent (vegetative shoots composed of leaves
    only); perigynium teeth ≤ 1 mm long, a third very small tooth
    often present . . . . 86. *C. lacustris*
    14. Vegetative culms well developed, in some species overtopping
    the fertile culms; perigynium teeth typically > 1 mm long, no
    third tooth present (section *Carex*)
        15. Perigynia pubescent
            16. Inner band of the uppermost leaf sheaths red to purple
            and thickened at the summit, glabrous; native, in wetlands
            throughout southern Wisconsin . . . . 90. *C. trichocarpa*
                                          (in part)
            16. Inner band of leaf sheaths not colored, pubescent;
            adventive at two southern Wisconsin sites . . . . 91. *C. hirta*
        15. Perigynia glabrous
            17. Inner band of the uppermost leaf sheaths red to purple
            and thickened at the summit, glabrous . . . . 90. *C. trichocarpa*
                                          (in part)
            17. Inner band of leaf sheaths pale or brown, not thickened
            at the summit, glabrous or pubescent
                18. Vegetative culms hollow, easily flattened; inner
                band of the leaf sheaths pubescent, rarely glabrous,
                not obviously veined; basal leaf sheaths ladder
                fibrillose . . . . 92. *C. atherodes*
                18. Vegetative culms solid; inner band of the leaf
                sheaths strongly veined, glabrous or the veins
                scabrous; upper and lower leaf sheaths ladder
                fibrillose . . . . 93. *C. laeviconica*
7. Perigynia > 10 mm long; spikes globose to oblong

19. Perigynia lanceolate, somewhat inflated, becoming yellowish
green, mostly < 3 mm wide; basal leaf sheaths green to yellow
or brown, never red or purple (section *Rostrales* [= *Folliculatae*])
   20. Leaf blades ≥ 3.5 mm wide; bract sheaths prolonged at the
summit, convex or truncate; central Wisconsin . . . . 94. *C. folliculata*
   20. Leaf blades ≤ 3.5 mm wide; bract sheaths not prolonged at
the summit, concave; Ashland and Bayfield counties . . . . . 95. *C. michauxiana*
19. Perigynia ovoid, strongly inflated, dull green and often becoming
brown, mostly > 3 mm thick; basal leaf sheaths reddish to purple,
sometimes brown (section *Lupulinae*)
   21. Pistillate spikes approximately as long as wide; perigynia
2–35 per spike; sheath of the uppermost leaf (not the bracts)
≤ 1.5 cm long (occasionally to 2.5 cm)
      22. Perigynia erect to spreading or the lower reflexed, lustrous,
glabrous; mesic to wet forests . . . . 96. *C. intumescens*
      22. Perigynia radiating in all directions from a central point,
dull or pubescent; floodplain forests . . . . 97. *C. grayi*
   21. Pistillate spikes longer than thick; perigynia 8–90 per spike;
sheath of the uppermost leaf (not the bracts) usually
≥ 1.7 cm long
      23. Achenes typically ≤ 2.6 mm wide, the angles not pointed
or knobbed; common in wet forests throughout the
state . . . . 98. *C. lupulina*
      23. Achenes typically ≥ 2.4 mm wide, the angles pointed or
knobbed; rare in southern Wisconsin wet forests . . . . . . 99. *C. lupuliformis*

Key to *Carex* Subgenus *Vignea*

The distinction between an androgynous spike and a gynecandrous spike can be difficult to see at first. The lower scales on a gynecandrous spike hold staminate flowers, which are obvious at flowering time by their dangling stamens. After the stamens fall off, the bare filaments are visible for at least a short time. When the perigynia ripen, the gynecandrous condition is often evident only by the presence of apparently empty scales at the base of the spike. These sometimes form a prominent clavate base, but in many species only a few scales will be visible. An androgynous spike has staminate flowers at its tip, and the same phenological sequence applies: at flowering time, look for stamens; after flowering is past, look for filaments or empty scales. The tiny cluster of staminate flowers at the tip of an androgynous spike is often hidden by the perigynium tips, especially once the stamens fall and the perigynia ripen.

To save yourself time and frustration, look first for staminate flowers (or apparently empty scales) at the base of an otherwise pistillate spike. If you find any, the spike is almost certainly gynecandrous. Very occasional plants have staminate flowers at both the base and the tip of the spike. If you don't find staminate scales at the base, scrounge around in the tips of a few spikes to see whether there are staminate flowers (or apparently empty scales) lurking amongst the perigynium beaks. You'll soon learn to recognize gynecandrous spikes quickly in the field.

I. Spikes androgynous or unisexual, or inflorescence mixed (some mix of
   androgynous, staminate, and/or pistillate spikes)
   II. Rhizomes long, conspicuous, often creeping; shoots typically arising
      singly or few to a clump . . . . Key F
   II. Rhizomes short and/or inconspicuous, though vegetative shoots may
      be elongate and behave like runners or stolons; shoots typically
      cespitose
      III. Inflorescence compound, at least the lower 1–2 spikes branching . . . . Key G
      III. Inflorescence simple, no spikes branching . . . . Key H
I. Spikes gynecandrous or inflorescence mixed (some mix of gynecandrous,
   staminate, and/or pistillate spikes)

IV. Perigynium margins not conspicuously flattened, generally rounded or
acute, ≤ 0.1 mm wide; vegetative culms lacking . . . . Key I
IV. Perigynium margins conspicuously flattened (winged), widest portion
≥ 0.2 mm wide; vegetative culms present though often inconspicuous,
at least until later in the season (section *Ovales*) . . . . Key J

KEY F—ANDROGYNOUS OR UNISEXUAL *VIGNEA* SPECIES WITH LONG, CONSPICUOUS RHIZOMES

All sedges have rhizomes, which may range from short and often inconspicuous to long and creeping. The species in this key have elongate rhizomes;
species that reproduce by stolons but not by rhizomes are not in this key.
Prominently stoloniferous species in our flora include *Carex assiniboinensis*
[43] and *C. chordorrhiza* [112]. Additionally, *C. projecta* [139], *C. tribuloides*
[138], *C. longii* [146], and *C. limosa* [19] reproduce by sending up shoots
from the nodes of fallen vegetative culms.

1. Inflorescence unisexual
  2. Spikes several, aggregated into a dense, elongate inflorescence 1–4 cm
    long; perigynia (on pistillate plants) ascending; leaf blades 1–3 mm
    wide; highway rights-of-way and occasional barrens, roadsides and
    other dry open habitats, scattered in northeastern and southeastern
    Wisconsin (section *Divisae*) . . . . 100. *C. praegracilis*
  2. Spike solitary, terminal, 0.5–1.5 cm long; perigynia (on pistillate plants)
    divergent or reflexed at maturity; leaf blades ≤ 1 mm wide; calcareous
    wetlands, primarily northeastern quarter of Wisconsin (section
    Physoglochin) . . . . 101. *C. gynocrates* (in part)
1. Inflorescence bisexual
  3. Rhizomes slender, threadlike, smooth; perigynia divergent at maturity,
    margins rounded (section *Dispermae*) . . . . 123. *C. disperma*
  3. Rhizomes thick, cordlike, typically scaly; spikes many-flowered; perigynia
    ascending at maturity, margins acute to narrowly winged, serrulate
    4. Inner band of the leaf sheaths strongly green-nerved; inflorescence
      2.5–7 cm long, often arching; lower spikes separate; wetlands (section
      Holarrhenae) . . . . 102. *C. sartwellii*
    4. Inner band of the leaf sheaths whitish-hyaline; inflorescence 1–5 cm
      long, typically straight; spikes all overlapping; dry sandy uplands
      and occasional wet prairies (section *Ammoglochin*) . . . . 103. *C. siccata*

KEY G—COMPOUND-SPIKED *VIGNEA* SECTIONS; SECTIONS *VULPINAE, MULTIFLORAE, HELEOGLOCHIN*, AND *PHAESTOGLOCHIN* (IN PART)

The species in this key have branched lower spikes. A sedge inflorescence with
branched spikes is referred to as compound or paniculate. (The treatments

in *FNA* use both of these terms; the latter applies only to inflorescences, while the former applies to any structure that comprises two or more structural entities that are equivalent to one another.) Learn how to recognize a compound inflorescence by dissecting a few spikes, removing the scales and perigynia to expose the underlying branches. It probably won't be necessary to do this more than a few times to get a sense for what a compound inflorescence looks like. Typically, a spike has lobes corresponding to the branches and is readily identified as compound or simple in the field.

Theodore S. Cochrane, in his treatment of section *Heleoglochin* for *FNA*, notes that sections *Vulpinae, Multiflorae,* and *Heleoglochin* share an additional, distinctive character: short, slender, tough, dark or even blackish rhizomes. Intact leaf sheaths provide essential information for identifying many species in the group.

1. Culm ≤ 2 mm wide at the base (make sure you measure the culm, not the leaf sheath), firm, angles not winged; leaf blades generally < 5 mm wide
   2. Inner band of the leaf sheaths corrugated (*Multiflorae*)
      3. Leaves longer than the culms; perigynia green, gradually narrowed to a beak half to one-third the total perigynium length . . . . 104. *C. vulpinoidea*
      3. Leaves shorter than the culms; perigynia yellowish, more abruptly narrowed to a beak ≤ one-quarter the total perigynium length . . . . . 105. *C. annectens*
   2. Inner band of the leaf sheaths not corrugated (*Heleoglochin*)
      4. Inner band of the leaf sheaths hyaline, red-dotted (10× magnification); perigynia 2–2.5 (rarely 3) mm long, lustrous, dark olive to brown at maturity, unequally biconvex, the back longitudinally grooved the full length of the body . . . . 106. *C. diandra*
      4. Inner band of the leaf sheaths strongly copper-colored at the apex, often with some red dots; perigynia 2.3–3 mm long, dull, light to dark brown at maturity, planoconvex, the back not grooved or grooved just at the base . . . . 107. *C. prairea*
1. Culm ≥ 2 mm wide at the base, spongy, easily compressed between the fingers, angles often scabrous and narrowly winged; leaf blades frequently > 5 mm wide
   5. Perigynia widest near the middle, weakly if at all spongy at the base (*Phaestoglochin* in part—go to Key H, couplet 4)
   5. Perigynia widest near the base, thick-spongy at the base (*Vulpinae*)
      6. Inner band of the leaf sheaths strongly corrugated; perigynia triangular in outline, 4–5 mm long, tapering to the apex, thick-spongy at the base; common in wetlands statewide . . . . 108. *C. stipata*
      6. Inner band of the leaf sheaths weakly if at all corrugated; perigynia various; not as common as *Carex stipata* [108]
         7. Perigynium base abruptly spongy-thickened, significantly thicker than and clearly distinguished from the rest of the body; perigynium beak narrow with nearly parallel sides, significantly longer than the body; uncommon in ditches, southeastern Wisconsin . . . . 109. *C. crus-corvi*
         7. Perigynium base spongy or not, in either case continuous with the rest of the body; perigynium beak tapering, shorter than the body; wetlands, southern Wisconsin

8. Inner band of the leaf sheaths thickened at the summit, neither corrugated nor dotted; perigynia obviously spongy-based; rare in Dane, Iowa, and Monroe counties . . . . 110. *C. laevivaginata*

8. Inner band of the leaf sheaths thin and fragile at the summit, rarely corrugated, typically purple-dotted; perigynia not spongy-based; more widespread, primarily in the southern quarter of the state and counties adjacent to Lake Winnebago . . . . 111. *C. alopecoidea* (in part)

KEY H—ANDROGYNOUS OR STAMINATE *VIGNEA* SECTIONS WITH SIMPLE SPIKES AND SHORT AND/OR INCONSPICUOUS RHIZOMES

1. Terminal spike staminate, though some plants in the population may have gynecandrous or pistillate terminal spikes (*Stellulatae* in part)

  2. Spike solitary; plants typically bisexual, occasionally unisexual; leaf blades involute, ≤ 1.5 mm wide . . . . 131. *C. exilis* (in part)

  2. Spikes ≥ 3; plants almost always unisexual; leaf blades flat or folded, 1–2.5 mm wide . . . . 132. *C. sterilis* (in part)

1. Terminal spike androgynous

  2. Perigynia biconvex, the margins rounded; plants long-stoloniferous, the prostrate stems producing upright culms at the nodes; inflorescence capitate, spikes congested; bogs and other sphagnous wetlands (*Chordorrhizae*) . . . . 112. *C. chordorrhiza*

  2. Perigynia planoconvex to biconvex, the margins acute to narrowly winged; plants not stoloniferous; inflorescence capitate to elongate, spikes congested to distant; dry to mesic habitats (*Phaestoglochin*)

    3. Lower leaf sheaths loose, the areas between the veins on the back conspicuously pale or whitened with conspicuous cross-veins (10× magnification); widest leaf blades mostly > 4 mm (occasionally 3 mm) wide; inflorescence compound or simple

      4. Spikes distant, the internode between the two lowest spikes > 10 mm, often > 20 mm . . . . 113. *C. sparganioides*

      4. Spikes overlapping to congested, the internode between the two lowest spikes usually ≤ 10 mm

        5. Inner band of leaf sheaths smooth, conspicuously red-spotted (10× magnification); culm weak, easily compressed; achenes 1.5 mm long, 1.3 mm wide (*Vulpinae* in part) . . . . . 111. *C. alopecoidea* (in part)

        5. Inner band of leaf sheaths usually smooth, not red-spotted (occasionally corrugated in *Carex muehlenbergii* [119]); culm firm; achenes 1.6–2 mm long, 1.3–2 mm wide

          6. Pistillate scale apex obtuse to acuminate or short-awned; scale body concealing ≤ half of the perigynium; perigynia green to pale brown; achene 1.6–2 mm long, 1.3–1.7 mm wide; mesic forests and woodland edges . . . . 114. *C. cephaloidea*

          6. Pistillate scales awned or acuminate; scale body concealing > half of the perigynium; perigynia typically becoming dark brown at maturity; achene 1.8–2.1 mm long, 1.6–2 mm wide; dry open areas, mostly disturbed soils

            7. Inner band of the leaf sheaths thin, fragile at the apex; backs not white-mottled; widespread, southern Wisconsin . . . . 115. *C. gravida*

7. Inner band of the leaf sheaths thick, firm at the apex; backs
often white-mottled; known in Wisconsin from a single
1979 collection in Lafayette County . . . . 116. *C. aggregata*
3. Lower leaf sheaths tight, the areas between the veins on the back
green to pale or whitened, lacking conspicuous cross-veins; widest
leaf blades ≤ 4 mm (occasionally 5 mm) wide; inflorescence simple
  8. Inflorescence open, the lowest internode many times as long as the
lowest spike
    9. Stigmas thick, tightly coiled; culms 0.5–1 mm wide beneath the
inflorescence, leaning or erect at maturity . . . . 117. *C. rosea*
    9. Stigmas narrow, straight or loosely curled; culms ≤ 0.5 mm wide
beneath the inflorescence, leaning or sprawling at maturity . . . . . . 118. *C. radiata*
  8. Inflorescence congested, the lowest internode ≤ 2 times as long as
the lowest spike
    10. Pistillate scales ≥ two-thirds as long as the perigynia
      11. Ligules ≤ 3 mm long; inner band of the leaf sheaths usually
smooth, occasionally corrugated, yellowish and thickened
at the summit, otherwise hyaline; basal leaf sheaths pale to
brown or black; widespread, southern half of Wisconsin
and Door County . . . . 119. *C. muehlenbergii*
      11. Ligules 4–8 mm long; inner band of the leaf sheaths smooth,
hyaline, not thickened at the summit; basal leaf sheaths
reddish to purplish, often very subtly; known from few
collections in central and south-central Wisconsin . . . . 120. *C. spicata*
    10. Pistillate scales < two-thirds as long as the perigynia
      12. Pistillate scales acuminate to toothed at the apex; inner band
of the leaf sheath thickened at the apex (though this can be
very difficult to see, especially in the field); perigynia ≥ half
as wide as long, widest at or near the middle of the body;
common, native in fields, woodlands, and forests . . . . . 121. *C. cephalophora*
      12. Pistillate scales acute to short-toothed; inner band of the
leaf sheath not thickened at the apex; perigynia typically
≤ two-fifths as wide as long, widest near the base; uncommon,
adventive in lawns and gardens . . . . 122. *C. leavenworthii*

KEY I—GYNECANDROUS *VIGNEA* SECTIONS, PERIGYNIA WINGLESS TO VERY NARROWLY WINGED

1. Perigynium beak apex entire or very weakly bidentate (beak absent in *Carex
tenuiflora* [124]); perigynium margins generally rounded, smooth to finely
serrulate (serrate in *C. arcta* [126]) (*Glareosae*)
  2. Perigynia beakless, obscurely veined; spikes approximately as long as
wide; restricted to far northern Wisconsin and Cedarburg Bog (Ozaukee
County) . . . . 124. *C. tenuiflora*
  2. Perigynia beaked, distinctly veined in most species; spikes mostly longer
than wide; more widespread
    3. Spikes 1–3, widely separate; perigynia few per spike; lowest bract
nearly equaling to overtopping the inflorescence . . . . 125. *C. trisperma*
    3. Spikes > 3, the upper typically overlapping; perigynia > 5 per spike;
lowest bract shorter than the inflorescence, often shorter even than the
spike it subtends

4. Perigynia widest near the base, beak margins conspicuously serrate
(10× magnification); spikes overlapping, often indistinct, forming a
dense terminal cluster . . . . 126. *C. arcta* (in part)
  4. Perigynia widest near the middle, beak margins smooth or finely
  serrate; spikes small, distinct, at least the lowest separate
    5. Perigynia ascending, 5–10 per spike, no suture (seam) visible on
    the back (10× magnification); spikes generally longer than
    wide . . . . 127. *C. canescens*
    5. Perigynia spreading, 1–5 per spike, suture (seam) visible on the
    back, especially on the beak (10× magnification); spikes
    approximately as long as wide . . . . 128. *C. brunnescens*
1. Perigynium beak apex conspicuously bidentate; perigynium margins flat
or rounded, finely to conspicuously serrate
  6. Lower perigynia appressed or ascending at maturity; perigynia widest
  near the middle (*Deweyanae*)
    7. Perigynia < 4 times as long as wide, ≥ 1.3 mm wide, veinless or weakly
    veined on the back . . . . 129. *C. deweyana*
    7. Perigynia > 4 times as long as wide, ≤ 1.3 mm wide, strongly veined
    on the back . . . . 130. *C. bromoides*
  6. Lower perigynia spreading or divergent at maturity; perigynia widest
  near the base
    8. Spikes 5–15, the upper densely congested and often indistinct from
    one another, bases not obviously clavate (*Glareosae* in part) . . . . 126. *C. arcta* (in part)
    8. Spikes 2–8, all distinct from one another, base of the terminal spike
    (if pistillate) prominently clavate (*Stellulatae* in part)
      9. Terminal spike staminate or, rarely, gynecandrous, the staminate
      portion < 1 mm long, base not prominently clavate . . . . . . 132. *C. sterilis* (in part)
      9. Terminal spike gynecandrous, staminate portion often > 1 mm
      long, base prominently clavate
        10. Lowest perigynia mostly 2–3 mm long, 1–2 times as long as
        wide, beak typically 0.5–1 mm long, < half as long as the
        body; calcareous wetlands, primarily in the eastern half of
        the state . . . . 133. *C. interior*
        10. Lowest perigynia mostly 3–4.75 mm long, 1.8–3.6 times as
        long as wide, beak typically 1–2 mm long, ≥ half as long as the
        body; primarily in acidic wetlands in the northern half and
        central sand counties of Wisconsin . . . . 134. *C. echinata*

KEY J—SECTION *OVALES*

Section *Ovales* comprises approximately 15 percent of North American
Carex species and is the largest section in subgenus *Vignea*. The section is
for the most part limited to North America, but it is ecologically widespread.
In the Midwest you will find members of this section in dry to mesic prai-
ries (*Carex brevior* [151]), wet prairies (*C. bebbii* [154], *C. bicknellii* [147],
C. festucacea [159], *C. molesta* [149], *C. scoparia* [153], *C. tenera* var. *tenera*
[157a]), floodplain forests and alluvial wetlands (*C. muskingumensis* [136],
C. projecta [139], *C. tribuloides* [138], *C. cristatella* [137], *C. tenera* var.

echinodes [157b]), and wet sand flats (*C. crawfordii* [152], *C. cumulata* [145], *C. straminea* [143], *C. sychnocephala* [135]). Plants in section *Ovales* tend to inhabit disturbed sites and produce viable seeds in abundance.

The section was divided into eleven informal species groups by Kenneth Kent Mackenzie, a lawyer working in the early twentieth century whose monograph of North American *Carex* is one of the most substantial contributions to our understanding of the genus. The species groups he named within *Ovales* are for the most part unnatural, meaning that they do not represent discrete evolutionary lineages. You can learn the section without ever knowing Mackenzie's subsections (and, in fact, his system was not followed in the *FNA* treatment of section *Ovales*). Learning to recognize a few species groupings, however, may help in learning the section, just as learning sections is helpful in learning the entire genus *Carex*. The following groups for the most part follow Mackenzie's system but without his informal (and never properly published) names.

The "Sychnocephala" group [135] is made up of one New World species and one Old World species, both of which are distinguished from the rest of the section by having *foliose bracts more than three times as long as the inflorescence* and *perigynia exceedingly long-tapering, the beak often twice as long as the body, with nearly parallel margins*. These two species are sufficiently distinct from the remainder of the section that they have often been segregated into a separate section, *Carex* section *Cyperoideae*. However, molecular genetic evidence does not support this separation.

The "Tribuloides" group [136–139] is confined to eastern North America, primarily in floodplain forests. All four species of the group are found in Wisconsin. Recognize this group by the *elongated and leafy vegetative culms*, which serve in vegetative reproduction in *Carex projecta* [139], *C. tribuloides* [138], and perhaps *C. cristatella* [137]. These species also have *thin, scalelike, narrowly winged perigynia, the wing often not extending to the base of the perigynium*. Other species in the section with prominent vegetative culms (e.g., *C. longii* [146]) have leaves clustered near the tops of the vegetative culms, whereas species in the "Tribuloides" group have leaves more evenly spaced along the upper half of the vegetative culms.

The "Foenea" group [140–142], more common in western North America than in Wisconsin, is united in the Wisconsin flora by a single but distinctive characteristic: *pistillate scales conceal the perigynia at maturity*. In interpreting this character, make sure you are looking at plants with mature perigynia: perigynia in all sedges are concealed by the scales when very immature.

The "Suberecta" group [143–146] has *perigynium bodies broadest near or above the midpoint* and *inner band of the leaf sheaths green-veined.* Most members of the group inhabit wet sandy soils.

The "Festucacea" group [147–157] makes up the core of section *Ovales* in eastern North America. Its species are characterized by a combination of characters: *inner band of the leaf sheaths mostly whitish-hyaline; inflorescence bracts not typically exceeding the inflorescence in length; pistillate scales not concealing the perigynia;* and *perigynia planoconvex, winged to the base, and widest below the midpoint.* This is something of a "trash basket" group, best described as the species that don't fit into any of the other species groups.

1. Longest bract ≥ 3 times as long as the inflorescence, foliose; perigynium beak longer than the body (often 2 times as long); inflorescence densely headlike; spikes longer than wide ("Sychnocephala" group) . . . . 135. *C. sychnocephala*
1. Longest bract shorter than to approximately 2 times as long as the inflorescence (rarely 3 times as long), slender, not foliose (the longest bracts rarely foliose); perigynium beak shorter than to as long as the perigynium body; inflorescence and spikes various
  2. Vegetative culms prominent, often more numerous than the flowering culms, leaves evenly spaced on the upper half; perigynia thin, flat, scalelike; floodplains ("Tribuloides" group)
    3. Leaf sheaths tight to the summit, margins rounded, inner band green-veined nearly to the summit; vegetative culms stiff, canelike, reminiscent of a long-leaved *Dulichium,* not producing young shoots at the nodes; spikes > 1 cm long, > 3 times as long as wide, narrowly tapered at both ends; perigynia 6–9 mm long, 2–2.5 mm wide . . . . 136. *C. muskingumensis*
    3. Leaf sheaths loose, expanded near the summit, margins sharply angled, inner band usually with an elongate, triangular hyaline region extending downward from the summit; vegetative culms weak, often becoming recumbent or prostrate, frequently producing young shoots from the nodes, in this regard suggesting vegetative reproduction in *Glyceria;* spikes < 2 times as long as wide, base tapered or rounded, apex rounded or obtuse; perigynia 2.5–5.5 mm, < 2 mm wide
      4. Spikes spherical at maturity; pistillate scales concealed by spreading perigynium tips . . . . 137. *C. cristatella*
      4. Spikes as long as to longer than wide, base tapered, apex round to obtuse; pistillate scales not concealed
        5. Inflorescence straight; spikes overlapping; perigynia > 30 per spike, tips ascending or appressed; southern two-thirds of the state . . . . 138. *C. tribuloides*
        5. Inflorescence arching or nodding; lowest spikes usually separate; perigynia 15–30 per spike, tips spreading at maturity; predominantly in the northern two-thirds of the state . . . . 139. *C. projecta*
  2. Vegetative culms generally inconspicuous, not as numerous as the flowering culms, at least until late in the season, leaves clustered near the summit; perigynia planoconvex, rarely scalelike; habitats various, rarely in floodplains (except *Carex tenera* var. *echinodes* [157b])

6. Pistillate scales equaling or exceeding the tips of the ripe perigynia, generally concealing the beaks ("Foenea" group)

  7. Pistillate scales as wide as the perigynia, concealing them; perigynium beaks flattened, margins finely serrate; inflorescence stiff; spikes mostly or all overlapping . . . . 140. *C. adusta*

  7. Pistillate scales narrower than the perigynia, covering the perigynium beak but not the body; inflorescences various

    8. Perigynium beaks flat in cross-section, margins finely serrate to the tip; inflorescence arching to nodding; native, northern Wisconsin . . . . 141. *C. foenea*

    8. Perigynium beaks round in cross-section, margins smooth at the tip (endmost 0.5 mm); inflorescence stiff; Eurasian, introduced in the Apostle Islands . . . . 142. *C. ovalis*

6. Pistillate scales shorter than the ripe perigynia, not concealing the beaks

  9. Perigynium bodies widest near or above the middle, usually obovate or diamond-shaped (if orbiculate—*Carex straminea* [143]—then achenes ≤ 1 mm wide, perigynia 4–5.5 mm long); inner band of the leaf sheaths green-veined nearly to the summit, sometimes with a narrow, tapering, hyaline region at the summit ("Suberecta" group)

    10. Inflorescence arching or nodding; perigynium body orbiculate, rounded at the base; uncommon in disturbed wet sand, primarily in Jackson and Monroe counties . . . . 143. *C. straminea*

    10. Inflorescence stiff; perigynium body obovate or diamond-shaped, wedge-shaped or rounded-acute to rounded-obtuse at the base

      11. Perigynium body approximately diamond-shaped in outline, wedge-shaped at the base, 4–5 mm long; distance from beak tip to achene 2–3 mm; rare in calcareous wetlands, southeastern Wisconsin . . . . 144. *C. suberecta*

      11. Perigynia obovate, rounded-obtuse to rounded-acute at the base, 3–4.5 mm long; distance from beak tip to achene 1.5–2.2 mm

        12. Inner face of the perigynia veinless; summit of the inner band of the leaf sheaths truncate; leaf blades 3–6 mm wide; open wet sands, generally disturbed, in Jackson County and adjacent counties . . . . 145. *C. cumulata*

        12. Inner face of the perigynia veined; summit of the inner band of the leaf sheaths concave; leaf blades 2–4.5 mm wide; known from a single LaCrosse County collection . . . . 146. *C. longii*

  9. Perigynium bodies usually ovate or elliptical (if orbiculate—in *Carex brevior* [151] and *C. festucacea* [150]—then achenes 1–2 mm wide, perigynia 2.5–5 mm long), not obovate or diamond-shaped, widest at or below the middle; inner band of the leaf sheaths hyaline or green ("Festucacea" group)

    13. Perigynia > 2 mm wide

      14. Inner face of the perigynia membranous, easily torn, translucent, revealing the dark brown achene; leaf sheaths papillose (20× magnification)

        15. Perigynium margins brown at maturity, contrasting with the body; inner perigynium face strongly veined over achene; largest perigynia > 5.5 mm long . . . . 147. *C. bicknellii*

15. Perigynium margins similar in color to the perigynium
　　body; inner perigynium face veined at the base,
　　sometimes weakly veined over the achene; largest
　　perigynia ≤ 5 mm long . . . . 148. *C. merritt-fernaldii*

14. Inner face of the perigynia firm, opaque, concealing the
　achene; leaf sheaths smooth, not papillose (except some
　Carex festucacea [150] individuals with papillose sheaths)

　　16. Spikes 2–4, congested, base and apex rounded; open
　　　wetlands, southern half of the state . . . . 149. *C. molesta*

　　16. Spikes mostly > 4, separate or overlapping, base tapering,
　　　apex acute or rounded; habitats and geographic
　　　distribution various

　　　　17. Inner face of the perigynia veined; spike bases
　　　　　clavate; inflorescence elongate, often arching or
　　　　　drooping; achenes 1–1.3 mm wide; primarily
　　　　　wet prairies . . . . 150. *C. festucacea* (in part)

　　　　17. Inner face of the perigynia veinless; spike bases
　　　　　tapered; inflorescence elongate, erect to arching,
　　　　　rarely drooping; achenes 1.3–1.8 mm wide; primarily
　　　　　dry prairies . . . . 151. *C. brevior*

13. Perigynia ≤ 2 mm wide

18. Perigynia > 2.5 times as long as wide, lanceolate

　　19. Spikes strongly overlapping, not always distinct,
　　　aggregated into a dense, headlike inflorescence of
　　　consistent form between populations; perigynia 0.9–
　　　1.3 mm wide . . . . 152. *C. crawfordii*

　　19. Spikes overlapping but distinct, forming a stiff or
　　　slightly arching inflorescence, the form of which is
　　　variable within and between populations; perigynia
　　　1.2–2 mm wide . . . . 153. *C. scoparia*

18. Perigynia < 2.5 times as long as wide, not lanceolate

　　20. Inflorescence compact, erect, all spikes overlapping,
　　　lowest inflorescence internode typically < 5 mm

　　　　21. Inner face of the perigynia veinless or faintly
　　　　　1–3 veined; inflorescence < 3 cm long . . . . 154. *C. bebbii*

　　　　21. Inner face of the perigynia typically 3-veined;
　　　　　inflorescence 1–6 cm long

　　　　　　22. Sheaths whitened between green veins on
　　　　　　　the backs, inner band not corrugated;
　　　　　　　perigynia green at maturity; throughout the
　　　　　　　state . . . . 155. *C. normalis* (in part)

　　　　　　22. Sheaths green between the veins on the backs,
　　　　　　　inner band often corrugated; perigynia brown
　　　　　　　at maturity; Apostle Islands . . . . 156. *C. tincta*

　　20. Inflorescence elongate, often arching or nodding, at
　　　least the lowest spikes separate, lowest inflorescence
　　　internode typically > 5 mm

　　　　23. Perigynium body orbiculate; leaf sheaths often
　　　　　papillose (20× magnification) . . . . 150. *C. festucacea* (in part)

　　　　23. Perigynia elliptical to ovate; leaf sheaths papillose
　　　　　only in *Carex tenera* var. *tenera* [157a]

24. Inflorescence erect, rarely arching; leaves 2–6.5 mm
    wide; sheaths not papillose . . . . 155. *C. normalis* (in part)
24. Inflorescence arching or nodding; leaves generally
    < 3.5 mm wide; sheaths papillose or not
    25. Perigynium beaks straight; sheaths papillose;
        prairies to upland oak forests . . . . 157a. *C. tenera*
        var. *tenera*
    25. Perigynium beaks spreading; sheaths not
        papillose; mesic and bottomland forests . . . . 157b. *C. tenera*
        var. *echinodes*

Carex Subgenus *Carex*

Bracts foliose in most species, sheathing the culm in many. **Cladoprophylls** *present (hidden by the bract sheaths).* **Lateral spikes** *usually pistillate or androgynous, if androgynous then generally with more pistillate than staminate flowers.* **Terminal spikes** *staminate or bisexual, the staminate flowers usually more abundant.* **Perigynia** *highly variable in shape, typically terete or triangular in cross-section, biconvex or planoconvex in cross-section in species with 2 stigmas.* **Achenes** *generally triangular in cross-section, biconvex in species with 2 stigmas.* **Stigmas** *usually 3, sometimes 2.*—Subgenus *Carex* is a large and morphologically heterogeneous group. In practice, the subgenus is easily recognized in the field because there is obvious division of sexes between the spikes of a given plant, each spike usually unisexual or dominated by either pistillate or staminate flowers. The subgenus is referred to as subgenus "Eucarex" in Fernald and several other treatments, but that name is invalid under the International Code of Botanical Nomenclature, which precludes naming infrageneric taxa using the prefix *Eu-*. The label "eucarex" (plural "eucarices") is used as an occasional convenience in this book to avoid the cumbersome phrase "members of subgenus *Carex.*"

Species descriptions in this portion of the book are abbreviated, generally focusing on characteristics that are most easily seen in the field and/or important for distinguishing similar species. Descriptions are especially short if the species has been described more fully in Part 2 of this book, in which case the page number for the full description is provided. Section descriptions are tailored to Wisconsin species and provided only for sections represented in the state by more than one species.

UNISPICATE "EUCARICES"—SECTIONS *LEUCOGLOCHIN*, *LEPTOCEPHALAE*, AND *PHYLLOSTACHYAE*

Spike solitary, terminal (basal lateral spikes sometimes present in C. backii *[2] and usually present in* C. jamesii *[1], in both cases typically inconspicuous).* **Stigmas** *3.*—See text under Key A (p. 29).

Section *Phyllostachyae* Tuckerman ex Kükenthal

Plants cespitose, bases brown. Bracts lacking. Lateral spikes absent or basal, pistillate or androgynous. Terminal spike androgynous. Lowest pistillate scale foliose, suggesting the lowest bract of the inflorescence in most other sections, exceeding the tip of the spike. Perigynia 2-ribbed, beak untoothed.— The foliose lower pistillate scale of these species is highly distinctive, shared only with section *Firmiculmes* of western North America.

    1. *Carex jamesii* Schweinitz. Plants strongly cespitose; leaves green, soft; lowest pistillate scales foliose, 1.5–3 mm wide, not concealing the perigynia; perigynium bodies globose, the beak abrupt, approximately as long as the body (p. 98).

    2. *Carex backii* W. Boott. Like *C. jamesii* [1], but lowest pistillate scales 2.5–6.5 mm wide, concealing the perigynia they subtend. Uncommon in dry to mesic forests of Dane County and Juneau County to LaCrosse County and Door County and far northeastern Wisconsin.

Section *Leucoglochin* Dumortier [= *Orthocerates* Koch]

The only other North American species in this section, *Carex microglochin* Wahlenberg, is common in northern Canada, Alaska, and the central Rocky Mountains.

    3. *Carex pauciflora* Lightfoot. Spike solitary, androgynous; perigynia 2–6, reflexed, lance-shaped (p. 99).

Section *Leptocephalae* L. H. Bailey [= *Polytrichoideae* (Tuckerman) Mackenzie]

This section contains only one species and is closely related to the genus *Uncinia*, which has been shown to be derived from a unispicate *Carex* ancestor.

    4. *Carex leptalea* Wahlenberg. Plants soft, very slender, rhizomatous; spike androgynous; perigynia few, appressed, finely and distinctly veined, often with a blunt or rounded, beakless apex (pp. 100–101).

Section *Hirtifoliae* Reznicek [formerly included in section *Triquetrae* (L. H. Bailey) Mackenzie]

    5. *Carex hirtifolia* Mackenzie. This species is easily recognized by the soft pubescence that covers the entire plant, including the distinctly beaked perigynia, which are 2-ribbed, otherwise veinless. Reminiscent of a softly pubescent *C. blanda* [50]. Plants loosely cespitose; rhizomes short; basal leaf sheaths pale to brown or green; leaf blades ≤ 8 mm wide, softly pubescent; perigynia 4–5 mm long, softly pubescent, beak ≥ 1 mm long, typically bent; stigmas 3. Rich forests throughout most of the state.

THE "PEACHFUZZ WOOD SEDGES"—
SECTIONS *CLANDESTINAE* AND *ACROCYSTIS*

Plants short, loosely cespitose or shoots arising singly (except Carex communis *[12], which is short-rhizomatous and often densely cespitose); bases mostly fibrous and/or colored. Rhizomes mostly slender, elongate, creeping. Leaf blades glabrous, in most species narrow. Spikes as long as thick or somewhat longer. Perigynia short-pubescent at least near the base of the beak but typically not so densely pubescent that surface features are hidden (often glabrous in* C. pedunculata *[6]). Stigmas 3. First sedges to flower each year, fruits maturing in April or May and falling soon after.*

Section *Clandestinae* G. Don [= *Digitatae* (Fries) H. Christ]

Basal sheaths not fibrous. Bracts reduced to bladeless sheaths. Perigynium beaks untoothed, generally < 0.5 mm.—Similar to section *Acrocystis* but readily distinguished by the characters listed.

    6. ***Carex pedunculata*** Mühlenberg ex Willdenow. Basal leaf sheaths red, bladeless; leaf tips brown, separated from the green blades by a red-purple line; lower spikes borne on slender, elongate stalks; perigynia pubescent or glabrous with an elongate base (pp. 102–103).

    7. ***Carex concinna*** R. Brown. The compact inflorescence of this species suggests some members of section *Acrocystis*, but the densely cespitose habit and coarse perigynium pubescence distinguish it. Plants densely cespitose; rhizomes short; basal leaf sheaths brown; widest leaves 2–3 mm; spikes congested near the tips of the culms, few-flowered, short; pistillate scales sparsely pubescent (may not be visible with only a hand lens); perigynia 2.5–3 mm long, densely pubescent with long, spreading hairs, short-beaked. Dolomitic wetlands of Door County and the Apostle Islands.

    8. ***Carex richardsonii*** R. Brown. Leaf blades pale green or yellow, often turning red with maturity; bract sheaths maroon, contrasting with the culms and foliage (pp. 104–105).

Section *Acrocystis* Dumortier [= *Montanae* Fries]

Basal leaf sheaths in most species become fibrous with age. Perigynium beaks bidentate, ≥ 0.5 mm long. Most common in dry woodlands, prairies, and open sand; less common in mesic woodlands or wet soils.—Similar to section *Clandestinae* but readily distinguished from it by the characters listed. This section contains a few notoriously difficult species groups, and identification generally requires making collections of whole plants with ripe perigynia. To evaluate whether perigynium bodies are orbicular (spherical) or elliptical (roughly football-shaped, longer than wide), collect ripe material and consider the silhouette of the perigynium body only, ignoring the beak and the base. Basal lateral spikes—spikes that arise near the base of the plant, often buried among the leaf bases—are key to identifying several species (*Carex deflexa* [9], *C. umbellata* [10], and *C. tonsa* [11]), and rhizomes and lower leaf sheaths are helpful in distinguishing taxa. Collect carefully, as the lower structures are easily lost.

9. *Carex deflexa* Hornemann. This northern plant has highly distinctive, plump, long-based perigynia that are not readily confused with anything else in the section. Plants loosely cespitose; basal leaf sheaths little if at all fibrous; culms on some plants all elongate, on others both short and long; perigynia ≤ 3 mm long, body globose, base long and narrow, beak 0.5–0.8 mm long. Sporadic in sandy or sphagnous, typically wet soils; mostly in the northern highlands with a few populations in the central sands. Wisconsin plants are *C. deflexa* var. *deflexa*.

10. *Carex umbellata* Schkuhr ex Willdenow [= *C. abdita* E. P. Bicknell]. Plants densely cespitose; basal leaf sheaths fibrous; at least some pistillate spikes inconspicuous, hidden among the leaf bases; perigynia ≤ 3.2 mm long, beak ≤ 1 mm long. Most common on calcareous prairies, especially dry (pp. 106–107).

11. *Carex tonsa* (Fernald) E. P. Bicknell. This species is similar to *C. umbellata* [10] but is sporadic in sandy and often acidic soils throughout the state, its perigynia longer (> 3 mm) with longer beaks. The two varieties have been considered separate species by some authors. *Carex tonsa* var. *tonsa* has pale green, relatively short, wide, smooth leaf blades and perigynia sparsely pubescent near the beak. *Carex tonsa* var. *rugosperma* (Mackenzie) Crins [= *C. rugosperma* Mackenzie] has bright green leaf blades that are much longer than the culms, softer, and usually somewhat scabrous on the surface and perigynia pubescent on all surfaces.

12. *Carex communis* L. H. Bailey. This is one of our most distinctive northern forest species, where the densely cespitose habit and relatively broad leaf blades make the plant recognizable even when it is not in fruit. Plants cespitose; rhizomes short; basal leaf sheaths red, generally not fibrous; widest leaf blades 3–7 mm wide; perigynia 2.5–4 mm long, bodies globose, beaks straight or slightly bent. Occasionally confused with *C. pensylvanica* [15], which differs in its long-rhizomatous habit, strongly fibrous basal leaf sheaths, and narrower leaf blades. Common in forests of the northern third of Wisconsin; occasional in mesic forests of southern Wisconsin and the Baraboo Hills. Wisconsin plants are *C. communis* var. *communis*.

13. *Carex lucorum* Willdenow ex Link. Similar to *C. pensylvanica* [15], but perigynium beaks > half as long as the body. Uncommon, Vilas, Oneida, and Marinette counties. Wisconsin plants are *C. lucorum* var. *lucorum*.

14. *Carex inops* L. H. Bailey. This plant is difficult to distinguish from the common *C. pensylvanica* [15], and its distribution in the Midwest may not be well understood. Like *C. pensylvanica*, except perigynia ≥ 1.5 mm wide, achenes > 2 mm long. Scattered woodlands and fields of the Driftless Area and northeastern Wisconsin, often in sandy soils. Wisconsin plants are *C. inops* ssp. *heliophila* (Mackenzie) Crins, which is recognized in older treatments as *C. heliophila* Mackenzie.

15. *Carex pensylvanica* Lamarck. Plants colonial, loosely cespitose or shoots arising singly; rhizomes elongate; basal leaf sheaths strongly fibrous; leaf blades abundant, long and fine, a mass of dead leaves often persisting at the base of the plant through winter; culms smooth or weakly scabrous just below the inflorescence; staminate spike 8–25 mm long; perigynia ≤ 3.2 mm long, ≤ 1.5 mm wide; achenes < 2 mm long. Pennsylvania sedge is the most widespread species of this section ecologically and geographically, growing in a wide range of dry to mesic woods and prairies (pp. 108–109).

16. *Carex peckii* Howe. Similar to *C. pensylvanica* [15], but perigynia yellowish; spikes congested near the apex of the culm, staminate spikes 5–9 mm long. Plants loosely cespitose; culms much taller than the leaves; perigynia yellowing at maturity, 3–4 mm long, not hidden by the scales, bodies ellipsoid. Open woodlands northward, most common in Door County.

17. *Carex novae-angliae* Schweinitz. Bill Crins, one of the foremost experts on section *Acrocystis*, notes that this short-rhizomatous plant forms discrete patches of light green foliage in forest understories. The plant is more common in Canada and the northeastern United States than in Wisconsin. Plants loosely cespitose; leaf blades ≤ 1.5 mm wide; lowest bract equaling or exceeding the tip of the inflorescence; lowest inflorescence internode > 7 mm; perigynia ≤ 2.5 mm long, concealed by the pistillate scales, bodies ellipsoid. Wisconsin special concern species, known from Ashland and Price counties.

18. *Carex albicans* Willdenow ex Sprengel. Similar to *C. peckii* [16], but scales nearly concealing the perigynia. Plants densely cespitose; rhizomes short; basal leaf sheaths little if at all fibrous; lowest spikes overlapping; perigynia 2–3.5 mm long, largely concealed by the pistillate scales, bodies ellipsoid. Sandy soils of central Wisconsin. *Carex albicans* var. *albicans* [=*C. artitecta*] has pistillate scales blunt or acute, the midrib not extending to the tip; it is known in Wisconsin from a dry rocky woodland atop a bluff in the Baraboo Hills. *Carex albicans* var. *emmonsii* (Dewey ex Torrey) Rettig [=C. *emmonsii*] has pistillate scales with a tiny bristle-tip formed by the end of the midrib. It is uncommon in wet sandy open woodlands of central Wisconsin.

Section *Limosae* (Heuffel) Meinshauser

Plants loosely cespitose or culms arising singly, strongly rhizomatous; bases reddish. Roots covered in a dense yellow feltlike tomentum. **Vegetative shoots** *becoming decumbent, behaving like stolons, producing shoots at the nodes.* **Pistillate spikes** *pendulous on slender stalks.* **Perigynia** *pale, short-beaked, papillose.* **Stigmas** *3.* — An easily recognized section, common in northern bogs and fens.

   19. *Carex limosa* Linnaeus. Leaf blades 1–2.5 mm wide, margins involute; pistillate spikes pendulous on slender stalks; pistillate scales as wide as the perigynia and concealing them (pp. 110–111).

   20. *Carex magellanica* Lamarck. Similar to *C. limosa* [19], but leaf blades flat, 1–4 mm wide, margins revolute; pistillate scales brown to blackish, narrower than the perigynia. North American plants are *C. magellanica* ssp. *irrigua* (Wahlenberg) Hultén [= *C. paupercula* Michaux].

Section *Porocystis* Dumortier [= *Virescentes* (Kunth) Mackenzie]

Plants cespitose. **Leaves and culms** *usually pubescent, at least sparsely.* **Pistillate** *spikes erect to spreading, ovoid to oblong-cylindrical.* **Perigynia** *beakless or very short-beaked, glabrous or pubescent.* **Stigmas** *3.* — Most species of this section are uncommon in Wisconsin; *Carex pallescens* [23] is common in far north-central Wisconsin. Cylindrical pistillate spikes of *C. pallescens* and *C. swanii* [21] resemble the spikes of *C. granularis* [54] and relatives.

   21. *Carex swanii* (Fernald) Mackenzie. The slender, cylindrical pistillate spikes and beakless, pubescent perigynia of this species are distinctive in the Wisconsin flora. Terminal spike gynecandrous; pistillate spikes erect, densely packed with perigynia, ≤ 4 mm thick; pistillate scales acuminate to short-awned, giving the otherwise tidy spikes a sparsely fringed appearance; perigynia compressed-triangular in cross-section, 1.5–2.5 mm long, beakless, pubescent with long, spreading hairs. Wisconsin special concern species, restricted to a few dry open sites in southeastern and central Wisconsin.

   22. *Carex bushii* Mackenzie. The flaring base of the gynecandrous terminal spike is reminiscent of *C. buxbaumii* [26]. Widest lateral spikes ≥ 4 mm thick; perigynia terete, beakless, glabrous or sparsely pubescent. Known from a single Wisconsin locality, a dry lime prairie in Iowa County, undoubtedly adventive.

23. *Carex pallescens* Linnaeus. Our only common member of the section, distinctive in the flora by its pubescent foliage, short-cylindrical pistillate spikes, and beakless, glabrous perigynia. Terminal spike staminate; pistillate spikes densely flowered, oblong to oblong-cylindrical, borne on threadlike stalks, the lowest erect or pendulous; perigynia terete, 2.5–3 mm long, beakless, weakly veined, glabrous. Wet ditches and occasional wet to mesic forests in far north-central and southeastern Wisconsin.

24. *Carex torreyi* Tuckerman. Similar to *C. pallescens* [23], but terminal spike rarely gynecandrous, and perigynia strongly many-veined, short-beaked. Very few collections in woodlands and savannas of Walworth, Waukesha, Trempealeau, and St. Croix counties.

Section *Anomalae* J. Carey

The only other North American species in this section grows in western North America.

25. *Carex scabrata* Schweinitz. Scabrous perigynia and upper leaf blade surfaces make this plant distinctive. Plants cespitose; basal leaf sheaths brown; leaf blades distinctly scabrous on edges and upper surfaces, glabrous; lowest bract sheathless; pistillate spikes 10–160-flowered, oblong or cylindrical, the lowest arching or pendulous on an elongate stalk; perigynia 3–4 mm long, distinctly veined, scabrous to sparsely short-pubescent, beak approximately 1 mm long and curved away from the inflorescence axis. Most common in wet woods and thickets; scattered in the central third of the state, locally common in the Apostle Islands and adjacent counties.

Section *Racemosae* G. Don [= *Atratae* (Heuffel) H. Christ]

Plants loosely to densely cespitose; rhizomes variable in length; bases dark red, generally fibrous; roots not clothed with yellow felt. Terminal spike gynecandrous (in our species). Pistillate scales dark, often black. Perigynia pale, often greenish, very short-beaked to beakless, smooth or papillose, 2-ribbed, inconspicuously veined (in our species). Stigmas 3.—The papillose surface texture of the perigynia is most easily seen with a microscope, but once learned it can be recognized in good light with a hand lens (10× magnification). This section is species-rich in western North America and northern Canada, much less so in the East.

26. *Carex buxbaumii* Wahlenberg. Spikes 3–5, loosely clustered or the lower ones separate; terminal spike typically gynecandrous; pistillate scales dark, narrow, long-pointed, at least some exceeding the perigynium tips; perigynia abruptly short-beaked to beakless (pp. 112–113).

27. Carex media R. Brown ex Richardson [= C. norvegica Retzius ssp. inferalpina (Wahlenberg) Hultén]. Perigynia pale and scales dark as in *C. buxbaumii* [26], but spikes usually 2–3, densely to loosely clustered; pistillate scales shorter than the perigynia; perigynia widest near the middle, short-beaked. Known from one population in Wisconsin, an algific talus slope in Grant County.

Section *Albae* (Ascherson & Graebner) Kükenthal

The only other North American member of this section is endemic to a single canyon in western Texas.

28. Carex eburnea W. Boott. Plants short, forming sods; rhizomes elongate; leaf blades involute, wiry; perigynia becoming black in age, beaks short, white-tipped (pp. 114–115).

Section *Paniceae* G. Don

Plants colonial, shoots arising singly or few together; rhizomes elongate; bases brown to maroon. Leaf blades typically stiff. Terminal spike staminate, typically raised above the uppermost pistillate spike. Lateral spikes generally cylindrical, ascending (except Carex vaginata *[29]). Perigynia several-veined, mostly short-beaked, papillose (except* C. vaginata*). Calciphiles, growing mostly in wet soils (but* C. meadii *[32] common in dry lime prairies).*—The section is fairly distinctive and easy to recognize in our flora. The singular *C. vaginata* [29] is morphologically distinct and may not be related to other members of the section.

29. Carex vaginata Tauscher. The nearly straight, tapering beaks of this species are unique among Wisconsin members of the section. Basal leaf sheaths pale brown; pistillate spikes loosely flowered, nodding, slender; perigynia 3.5–5 mm long, not papillose, beaks cylindrical, 0.5–2 mm long. White cedar swamps of northeastern Wisconsin and Douglas County.

30. Carex woodii Dewey. Lower leaf sheaths bladeless, red to purple; pistillate spikes loosely flowered (pp. 116–117).

31. Carex livida (Wahlenberg) Willdenow. Superficially similar to *C. meadii* [32] and *C. tetanica* [33], differing from these and Wisconsin's other members of section *Paniceae* in having strongly glaucous leaf blades and straight, spindle-shaped, beakless perigynia that are strongly ascending at maturity. Wisconsin special concern species known from bogs along the shore of Lake Superior and a handful of other bogs and fens in the state.

Wisconsin plants have been recognized as *C. livida* var. *radicaulis* Paine, but separate varieties are not recognized in the recent *FNA* treatment.

32. Carex meadii Dewey. Lower leaf sheaths brown; leaf blades generally grayish green, firm. Common in dry to dry-mesic prairies (pp. 118–119).

33. Carex tetanica Schkuhr. Very similar to *C. meadii* [32], but leaf blades typically green, softer; usually in wet prairies. See discussion under *C. meadii.*

Section *Ceratocystis* Dumortier [= *Extensae* Fries]

*Plants cespitose; rhizomes short; bases brown. **Terminal spike** staminate, occasionally androgynous. **Lateral spikes** pistillate, densely flowered, globose to oblong. **Perigynia** strongly veined, abruptly beaked; beak toothed, generally reflexed. **Stigmas** 3. Wet calcareous soils: see* Carex cryptolepis *[35] for range.*

34. Carex viridula Michaux. Similar to *C. cryptolepis* [35] and *C. flava* [36] in appearance and distribution, but perigynia spreading, rarely reflexed, beak straight or slightly reflexed, generally < half as long as the body.

35. Carex cryptolepis Mackenzie [= *C. flava* var. *fertilis* auct. non Peck]. Lower perigynia reflexed; perigynium beaks reflexed, > half as long as the body; pistillate scales the same color as the perigynia, mostly hidden by them (pp. 120–121).

36. Carex flava Linnaeus. Similar to *C. cryptolepis* [35] in appearance and distribution, but perigynium beaks finely scabrous near the tip (15× magnification) and pistillate scales brownish, conspicuous until the lower perigynia reflex.

THE "PENDULOUS WOOD SEDGES"—SECTIONS *HYMENOCHLAENAE* [= *SYLVATICAE*, *GRACILLIMAE*, AND *LONGIROSTRES*] AND *CHLOROSTACHYAE*

*Plants typically cespitose; rhizomes short. **Lowest bracts** leaflike; sheaths > 3 mm long (except* Carex castanea *[44], which has narrow, short-sheathing bracts). **Lateral spikes** pistillate, narrowly oblong to elongate-cylindrical, typically pendulous on slender, elongate stalks, the uppermost held well above the leaves. **Perigynia** mostly straight, glabrous; beaks present or absent, the apex entire or very short-toothed. **Stigmas** 3. Mostly species of mesic or wet forests and woodlands.* —Anyone who has botanized Wisconsin's maple and bottomland forests will recognize this group by its graceful, pendulous pistillate spikes. These sections were placed together into section *Hymenochlaenae* by Kükenthal (1909), who wrote the last worldwide treatment for the genus *Carex*, but together they represent a heterogeneous and unnatural group. Because it is not clear how these sections will be treated in the future, I refer to each of the traditional sections as a separate group within section *Hymenochlaenae*.

Section *Hymenochlaenae* (Drejer) L. H. Bailey—*Gracillimae* group

Basal leaf sheaths purple (brown in Carex prasina *[37]). Leaves glabrous or pubescent. Terminal spike typically gynecandrous (often staminate in* C. prasina*). Mostly in mesic or floodplain forests throughout the state.*

37. *Carex prasina* **Wahlenberg.** This plant is a striking member of our flora, distinguished by its flattened, arching perigynia and its disjunct distribution in the state. Basal leaf sheaths brown; leaves blue-green, sheaths glabrous; perigynia arched away from the inflorescence axis, sharply triangular in cross-section, tapering to a curved or bent beak, 3–4 mm long, 2-ribbed but otherwise veinless; terminal spike staminate or gynecandrous. A state threatened species of wooded seeps in the Baraboo Hills, on Rock Island (Door County), on the Apostle Islands, and in a few sites in northwestern Wisconsin.

38. *Carex gracillima* **Schweinitz.** Basal leaf sheaths purple; leaves dark green, sheaths glabrous; perigynia round-triangular in cross-section, tapered to the base, rounded at the apex, beakless; common in forests and woodlands throughout the state (pp. 122–123).

39. *Carex davisii* **Schweinitz & Torrey.** The large perigynia and prominent pistillate scale awns make this lovely plant highly distinct. Lower leaf sheaths pubescent; lowest bracts leaflike, generally exceeding the tip of the inflorescence; lateral spikes pistillate; pistillate scales prominently awned; perigynia 4.5–6 mm long, short-beaked. Floodplains in southwestern Wisconsin.

40. *Carex formosa* **Dewey.** The short-beaked perigynia subtended by scales that are at most short-bristle-tipped make this uncommon plant easy to identify. Leaf sheaths pubescent; lowest bracts shorter than to equaling the tip of the inflorescence; spikes usually all gynecandrous; pistillate scales acute or at most short-toothed at the apex; perigynia intermediate in size between those of *C. davisii* [39] and *C. gracillima* [38], short-beaked. State threatened species of mesic forests in southeastern Wisconsin and Brown, Calumet, Outagamie, and Marquette counties.

Section *Hymenochlaenae*—*Longirostres* group

41. *Carex sprengelii* **Dewey ex Sprengel [=** *C. longirostris* **Torrey].** Basal leaf sheaths gray-brown, strongly fibrous; perigynia 10–40 per spike, abruptly long-beaked (pp. 124–125).

Section *Chlorostachyae* Tuckerman ex Meinshausen
[= *Capillares* (Ascherson & Graebner) Rouy]

42. *Carex capillaris* Linnaeus. The fibrous basal leaf sheaths and small beadlike perigynia borne in slender spikes on threadlike stalks are distinctive. Plants diminutive, densely cespitose; basal leaf sheaths fibrous; leaf blades 1–4 (often ≤ 2) mm wide; culms fine; lateral spikes pendulous on capillary stalks; perigynia 4–20 per spike, dark and lustrous at maturity, 2–3 mm long, 2-ribbed, tapering to the beak and the base. Wisconsin special concern species, mostly restricted to lakeshore seeps and sphagnous wetlands in Door and Bayfield counties.

Section *Hymenochlaenae*—*Sylvaticae* group

Superficially similar to the *Gracillimae* group but more delicate overall. *Terminal spike wholly staminate. Perigynia 8–45 per spike (1–8 in* Carex assiniboinensis *[43]), typically narrow, tapering to elongate beaks. Woodlands and wetlands northward.*

43. *Carex assiniboinensis* W. Boott. The long-arching vegetative shoots of this species are unique in the genus. They elongate in mid- to late summer and produce plantlets at the tips, allowing the species to spread vegetatively in the manner of a spider plant. Plants cespitose; basal leaf sheaths red, short-bladed, weakly pinnate-fibrillose; vegetative culms elongate, arching, bearing short, reflexed leaves and plantlets at the shoot apex; pistillate spikes loosely flowered, sometimes pendulous; perigynia tapered to both ends, 5–6.5 mm long, slender, short-pubescent; orifice of perigynium beak oblique, the long stigmas consequently emerging from one side of the perigynium tip. State threatened species; scattered populations in northern wet forests, frequently in floodplains of small rivers.

44. *Carex castanea* Wahlenberg. Similar to *C. arctata* [45] and *C. debilis* [46], but leaves pubescent (check with a 10× hand lens if you have any doubt); pistillate spikes 1–2.5 cm long, 4–5 mm thick. Margins of wet to mesic coniferous forests and bogs, northeastern and far northern Wisconsin.

45. *Carex arctata* W. Boott. Leaf blades 3–10 mm wide; pistillate spikes 2.5–8 cm long, 3–4 mm thick; perigynia 3–5 mm long, constricted abruptly at the base to a short stipe, apex tapered to a long beak. Common in forests of northern Wisconsin and the Baraboo Hills (pp. 122–123).

46. *Carex debilis* Michaux. Similar to *C. arctata* [45] but more slender overall. Leaf blades mostly 3–5 (occasionally 7) mm wide; perigynia 5–6 mm long, tapering to a slender, acute base without a stipe. Common in wet forests and woodlands of central and northern Wisconsin. Wisconsin plants are *C. debilis* var. *rudgei* L. H. Bailey.

THE "WOOD SEDGES"—SECTIONS *LAXIFLORAE*, *CAREYANAE*, *GRISEAE*, AND *GRANULARES*

Plants cespitose; rhizomes inconspicuous (except Carex crawei *[53], which has culms arising singly or in small clumps from elongate rhizomes). Leaf blades generally > 4 mm wide, glabrous. Pistillate spikes mostly ascending, the lowest nodding in some species. Perigynia many-veined (except C.* leptonervia *[49], which has all but 2–3 veins inconspicuous), glabrous, beakless or short-beaked, the beak abruptly bent in many species. Stigmas 3. Most common in forests, some plants common in calcareous wetlands.—* These four sections are morphologically similar and probably closely related to one another. Sections *Laxiflorae* and *Careyanae* have not always been treated as separate, but they can be distinguished from one another once the characters are understood. In evaluating whether perigynium veins are impressed or raised, make sure you have ripe material, preferably fresh (impressed veins show on dried perigynia in section *Griseae* but not in section *Careyanae*). Inspect the perigynium with a 10× hand lens in bright light and look along the margin to see whether the epidermis of the perigynium dips in at each vein—the impressed condition—or whether the veins appear as raised lines.

Section *Laxiflorae* (Kunth) Mackenzie

Plants cespitose; bases pale to brown or occasionally reddish. Culms weak, ascending to decumbent, sharply triangular in cross-section, angles sometimes winged. Perigynia triangular in cross-section with rounded edges, 25–40-veined (except Carex leptonervia *[49]); base stipitate, sometimes long-tapering; beak (in our species) abrupt, short, bent toward the inflorescence axis (more or less straight in C.* laxiflora *[52] and C.* leptonervia *[49]). Forests.—* Species in this section can be difficult to distinguish from one another, especially in the southeastern United States. In Wisconsin you will find mostly *C. albursina* [47] and *C. blanda* [50] in the south, *C. leptonervia* [49] in the north.

47. Carex albursina E. Sheldon. Leaf blades 1–4 cm wide; bracts wide, concealing the spikes, sheath margins smooth; pistillate scales blunt or truncate at the apex; perigynia 3–4.5 mm long (pp. 126–127).

48. Carex gracilescens Steudel. Similar to *C. blanda* [50] and *C. laxiflora* [52], but basal leaf sheaths reddish or purplish; staminate spike long-stalked; perigynia approximately 3 mm long. Primarily in mesic forests in the easternmost counties of the state, Brown County and southward.

49. Carex leptonervia (Fernald) Fernald. Similar to *C. blanda* [50], but perigynia 2–3 mm long with 2–3 distinct veins, the others subtle or indistinct (becoming more pronounced in dry specimens). Nearly as common in northern forests as *C. blanda* [50] is in the south.

50. Carex blanda Dewey. Similar to *C. albursina* [47], but widest leaf blades ≤ 1 cm wide; bracts not concealing the spikes, angles of the sheath margins scabrous on the inner band; perigynia 2.5–4 mm long, 25–32-veined (pp. 128–129).

51. Carex ormostachya Wiegand. Similar to *C. laxiflora* [52], but perigynia 2–3.5 mm long, beaks < 0.5 mm. Basal leaf sheaths purple-tinged, often inconspicuously; bract sheath margins smooth or papillose; pistillate spikes loosely flowered, the lowest perigynia overlapping barely or not at all. Forests of Bayfield and Door counties and northeastern Wisconsin.

52. Carex laxiflora Lamarck. Similar to *C. blanda* [50], *C. ormostachya* [51], and *C. gracilescens* [48], but basal leaf sheaths brownish; bract sheath angles smooth; pistillate spikes loosely flowered, slender; perigynia 3–4 mm long, beaks 0.5–1.5 mm. Mostly in beech forests in counties adjoining Lake Michigan; a few populations inland in northeastern Wisconsin.

Section *Granulares* (O. Land) Mackenzie

Plants cespitose or shoots arising singly from elongate rhizomes. Pistillate spikes oblong to narrowly oblong, densely packed with perigynia. Pistillate scales and perigynia dotted or finely streaked with red. Perigynia ≥ 25 per pistillate spike; veins 25–40, raised.—The red pigment flecks are most easily seen in the epidermis of the pistillate scales. Remove a scale and set it against a light background, then inspect it closely with a 10× hand lens under direct sunlight. Once you have learned this character and correlated it with other characteristics of the section, you will be able to recognize the *Granulares* as a group without each time having to verify the presence of the pigmentation.

53. Carex crawei Dewey ex Torrey. Like *C. granularis* [54], but culms arising singly from elongate rhizomes; widest leaf blades < 5 mm; staminate spike long-stalked, raised above—often well above—the uppermost pistillate spike. Wet open calcareous habitats, especially fens and shores; eastern Wisconsin.

54. Carex granularis Muhlenberg ex Willdenow. Densely cespitose with short, inconspicuous rhizomes; widest leaf blades > 5 mm wide; staminate spike overlapped by the uppermost pistillate spike (pp. 130–131).

Section *Griseae* (L. H. Bailey) Kükenthal
[= *Oligocarpae* (Heuffel) Mackenzie]

Perigynia round or obtusely angled in cross-section, many-veined; veins impressed in both fresh and dried material.—Perigynia in *Carex conoidea* [55] and *C. grisea* [56] suggest a balloon that has been inflated with many circles of twine tied lengthwise around it. In fact, this is analogous to what happens during development, as the cells that form the veins constrain the epidermis of the growing perigynium.

55. *Carex conoidea* Schkuhr ex Willdenow. Staminate spike long-stalked, elevated above the upper pistillate spikes; perigynia slightly inflated, yellowish, ≤ 4 mm long, beakless, lustrous (pp. 132–133).

56. *Carex grisea* Wahlenberg [= *C. amphibola* Steudel var. *turgida* Fernald]. Staminate spike sessile, often partly hidden by the lateral pistillate spikes; perigynia slightly inflated, greenish, ≥ 4.5 mm long, beakless (pp. 134–135).

57. *Carex hitchcockiana* Dewey. The combination of impressed perigynium veins and bent, tapering beaks in this and the following species is distinctive in our flora. Plants densely cespitose; basal leaf sheaths brown, short-pubescent; perigynia tightly enclosing the achene, 4.5–5.5 mm long, beak strongly curved away from the inflorescence axis. Mesic forests of southern Wisconsin and Door County.

58. *Carex oligocarpa* Schkuhr ex Willdenow. Similar to *C. hitchcockiana* [57], but basal leaf sheaths red-purple, glabrous; perigynia 3.5–4.5 mm long, beak straight or slightly curved away from the inflorescence axis. Local in mesic forests in the southern quarter of Wisconsin.

Section *Careyanae* Tuckerman ex Kükenthal
[formerly included in section *Laxiflorae*]

Resembling the *Laxiflorae* in general appearance, but *culms generally firm. Lowest pistillate spike typically basal. Perigynia acutely angled, tightly enclosing the achene; veins > 40, impressed in fresh material, raised in dried material.*—This section is best known in Wisconsin from the beautiful purple sheaths and evergreen foliage of plantain-leaved sedge (*Carex plantaginea* [59]).

59. *Carex plantaginea* Lamarck. The broad, dark green leaf blades and purple bract sheaths of this species are utterly distinctive. Basal leaf sheaths, bract sheaths, and staminate scales purple; bract sheaths short-bladed; leaf blades puckered on the margins, dark green, W-shaped in cross-section, > 1.5 cm wide; perigynia ≥ 9 per spike, 3.5–5 mm long. Relatively widespread in northeastern Wisconsin mesic forests; sporadic in forests of northern and west-central Wisconsin.

60. *Carex careyana* **Torrey ex Dewey.** This species is distinguished in the section by its relatively large perigynia with tapered beaks. Basal leaf sheaths but not bracts purple; leaf blades > 1 cm wide; perigynia ≤ 9 per spike, the largest ≥ 5 mm long. Wisconsin state threatened species. Extremely local in rich forests of Polk, Vernon, and LaCrosse counties.

61. *Carex platyphylla* **J. Carey.** Uncommon in the state, this species is easily recognized by its strongly glaucous leaves and slender culms. Leaf blades grayish to blue-green, strongly glaucous, the widest > 1 cm wide; culms frequently < 0.5 mm wide; perigynia ≤ 9 per spike, 2.5–3 mm long. Sugar maple–beech forests in Door County.

62. *Carex laxiculmis* **Schweinitz.** Basal leaf sheaths brown; culms 0.1–0.5 m tall, lax; leaf blades glaucous (var. *laxiculmis*) to bright green (var. *copulata*), the widest blades mostly ≤ 9 mm (occasionally > 1 cm); lateral spikes pistillate or gynecandrous, loosely flowered, the lowest pendulous, borne on a slender stalk emerging from the lowest sheath on the culm; lowest scale of each pistillate spike sterile (empty) or subtending a staminate flower; perigynia 2.5–4 mm long (pp. 136–137).

63. *Carex digitalis* **Willdenow.** Similar to *C. laxiculmis* [62], but widest leaf blades 3–5 mm wide; lowest scale of pistillate spikes subtending a perigynium at maturity. Rich forests, known from a handful of populations in southeastern Wisconsin. Wisconsin plants are *C. digitalis* var. *digitalis*.

Section *Bicolores* (Tuckerman ex L. H. Bailey) Rouy

Plants short, colonial, loosely cespitose, shoots arising singly or few in a clump; rhizomes elongate; bases brown. **Terminal spike** *staminate or gynecandrous, hidden by the crowded lateral spikes.* **Perigynia** *plump, weakly veined; margins and apex rounded, beakless to short-beaked, orifice entire.* **Stigmas 2.** *Calciphiles that favor some disturbance.*—One of only two distigmatic sections in subgenus *Carex*.

64. *Carex aurea* **Nuttall.** Perigynia ascending to spreading, becoming orange at maturity, fleshy; terminal spike typically staminate; lateral spikes pistillate, loosely flowered. Moist open or shaded ground along the Great Lakes, occasionally to 100 km inland; rare in central Wisconsin (pp. 138–139).

65. *Carex garberi* **Fernald.** Like *C. aurea* [64], but perigynia appressed-ascending, dry, whitish; terminal spike typically gynecandrous; pistillate spikes more densely flowered. Calcareous wetlands of Door County.

Section *Phacocystis* Dumortier [= *Acutae* (J. Carey) H. Christ] [= *Cryptocarpae* (Tuckerman ex L. H. Bailey) Mackenzie]

Plants often cespitose; rhizomes short or long. Lower leaf sheaths brown to red, fibrous in some species. Terminal spike typically staminate, ascending. Lateral spikes pistillate or androgynous, ascending to nodding (Acutae) or drooping (Cryptocarpae), elongate. Perigynia biconvex with distinct marginal veins. Stigmas 2. —Almost ubiquitous in Wisconsin wetlands, ranging from floodplains and wet forests (*Carex crinita* [66], *C. gynandra* [67], and *C. emoryi* [73]) to sedge meadows (*C. stricta* [70]), wet prairies (*C. haydenii* [71]), bogs and marshes (*C. aquatilis* [68]), and wet roadsides and ditches. Most of these species are dominant in their respective habitats, in which they play crucial roles in peat formation and nutrient cycling. Although the old sections *Acutae* and *Cryptocarpae* intergrade, they are easily distinguished in Wisconsin: the *Cryptocarpae* are characterized by long, drooping inflorescences and scabrous-awned pistillate scales; the *Acutae* have ascending to arching inflorescences and unawned pistillate scales. The name *Phacocystis* comes from Greek terms for "lentil" and "bladder," referencing the lens-shaped perigynia.

Carex crinita group [= section *Cryptocarpae*]

Plants densely cespitose, forming large, fountainlike tussocks. Lateral spikes long-drooping. Pistillate scales 3-veined with a long, toothed awn. Achenes typically pinched or wrinkled on 1 side.

66. Carex crinita Lamarck. Lower leaf sheaths smooth; pistillate scale body notched or truncate at the apex; perigynium apex rounded or blunt, abruptly narrowed to a short beak < 0.5 mm long. Wisconsin plants are *C. crinita* var. *crinita* (pp. 140–141).

67. Carex gynandra Schweinitz [= *C. crinita* var. *gynandra*]. Similar to *C. crinita* [66] in morphology and distribution, but lower leaf sheaths roughly pubescent; pistillate scale body typically tapering at the apex, not notched or truncate; perigynium apex acute.

Carex stricta group [= section *Acutae*]

Plants typically colonial, some cespitose; rhizomes elongate. Lateral spikes erect or nodding. Pistillate scales acute or acuminate at the apex, unawned. Achenes symmetrical, not pinched or wrinkled. —Basal leaf sheaths are important in identifying these species and should be collected intact.

68. Carex aquatilis Wahlenberg. Similar to *C. stricta* [70] and *C. emoryi* [73], but plant bases and lower leaf sheaths slightly spongy, not fibrous; lowest bract exceeding the tip of the inflorescence; pistillate/androgynous spikes mostly > 4 mm thick, the thickest often ≥ 5 mm; perigynia 2–3.5 mm long, widest above the middle, veinless, rounded at the apex; achenes glossy; staminate spikes 2–4. Two intergrading varieties occur in Wisconsin: *C. aquatilis* var. *aquatilis*, which has pistillate scales reddish to purplish with a narrow, pale midvein; and *C. aquatilis* var. *substricta* Kükenthal, which has pistillate scales pale brown with a broad, pale midvein. Most common in sedge meadows, fens, bogs, and shores of eastern and northern Wisconsin; sometimes in standing water (pp. 142–145).

69. *Carex lenticularis* Michaux. Similar to *C. aquatilis* [68], but pistillate/androgynous spikes very narrow (mostly < 4 mm wide); perigynia veined; staminate spike solitary. Wet sandy shores and sandstone crevices of Lake Superior; known in Wisconsin only from the Apostle Islands and Vilas County.

70. *Carex stricta* Lamarck. Strongly cespitose, forming peaty hummocks in wet, usually open sites; basal leaf sheaths red, ladder-fibrillose; pistillate scales shorter than the perigynia; perigynia ovate, pale, not inflated, the apex acute or obtuse; achenes dull (pp. 142–145).

71. *Carex haydenii* Dewey. This common wet prairie species is similar to *C. stricta* [70], but it forms dense clumps instead of hummocks; pistillate scales distinctly longer than the perigynia; perigynia ellipsoid, slightly inflated, < 2.5 mm long, the apex rounded and very minutely beaked. Low prairies, sandy sedge meadows, and bottomland woods throughout much of the state except the eastern quarter (pp. 142–145).

72. *Carex nigra* (Linnaeus) Reichard. This species of northeastern North America, Europe, and West Asia barely sneaks into Wisconsin, where it is easily recognized by its dark-spotted perigynia 2–3.5 mm long and black pistillate scales. Lower leaf sheaths not fibrous. Known in Wisconsin only from wetlands in and around the city of Superior.

73. *Carex emoryi* Dewey ex Torrey. Similar to *C. stricta* [70] but not cespitose; basal leaf sheaths not ladder-fibrillose; inner band of the leaf sheaths convex at the summit, forming a membranous appendage; ligules truncate, unlike the slender, inverted V-shaped ligule of *C. stricta* [70]; perigynia 1.5–3 mm long, veined; achenes dull. Floodplains, mostly in the southern two-thirds of Wisconsin.

THE "BLADDER AND BOTTLEBRUSH SEDGES"—SECTIONS *CAREX*, *LUPULINAE*, *PALUDOSAE* (IN PART), *ROSTRALES*, *SQUARROSAE*, AND *VESICARIAE*

Plants cespitose or shoots arising singly; rhizomes various; bases typically purple or reddish, sometimes brown. **Leaf blades** *septate-nodulose in most* *species.* **Pistillate spikes** *erect to arching, the lowermost pendulous and/or drooping in some species, elongate in most.* **Perigynia** *> 2.5 mm long, usually inflated, beak pronounced (short only in* Carex lacustris *[86]).* **Style** *continuous with the achene apex persistent.* **Stigmas** *3.*—See text under Key E (pp. 40–41).

Section *Squarrosae* J. Carey

The obconic perigynia of this section are highly distinctive, widest at the apex and abruptly narrowed to the beak; perigynia in all of the other "bladder" and "bottlebrush" sections taper more gradually to the beak. *Carex typhina* [74] is the only one of our "bladder" and "bottlebrush" sedges to occasionally produce unispicate plants.

    74. *Carex typhina* **Michaux.** The plump spikes, the terminal of which is gynecandrous, are highly distinctive. Plants cespitose; rhizomes short; terminal spike gynecandrous; pistillate spikes plump, oblong to ellipsoid, ≥ 1 cm thick; perigynia ascending, 5.5–8 mm long, tapering to the base; apex truncate or rounded, beak abrupt, 2–3 mm long. Floodplain forests of the Wisconsin, Lower Black, and Chippewa rivers.

Section *Vesicariae* (Heuffel) J. Carey

The section as circumscribed in *FNA* includes the typical bottlebrush sedges of the former section *Pseudocypereae* (**Tuckerman**) **H. Christ** [75–79], which have *pistillate spikes densely packed with perigynia* and *pistillate scales with scabrous awns that emerge conspicuously from between the perigynia,* as well as the former section *Vesicariae* [80–85], characterized by *pistillate spikes often narrower, more elongate, less densely packed with perigynia in some species* and *pistillate scales for the most part awnless, hidden by the perigynia.* These two sections overlap morphologically and as a consequence are probably best treated as a single section, though most Wisconsin species can for the most part be reliably placed in one of the two sections.

    75. *Carex comosa* **W. Boott.** Plants cespitose, often densely so; leaf blades W-shaped in cross-section, ≤ 17 mm wide; pistillate spikes ≥ 12 mm thick; perigynia reflexed, scarcely inflated, tightly enclosing the achene, beak teeth curved outward, 1.2–2 mm long (pp. 146–147).

    76. *Carex pseudocyperus* **Linnaeus.** Similar to *C. comosa* [75], but pistillate spikes narrower; perigynium beak teeth straight or only slightly curved, < 1.2 mm long. Lake margins, bogs, sedge meadows, and other wetlands in northern Wisconsin.

77. *Carex schweinitzii* Dewey ex Schweinitz. This plant is uncommon in the Midwest outside of Michigan. It is easily distinguished from other members of the section by its long-rhizomatous habit and staminate scale margins, which are smooth except at the very tip. Plants colonial, shoots arising singly from long rhizomes; culms weak, easily compressed; pistillate spikes slender, elongate; staminate scale margins smooth; perigynia loosely enveloping the achene, 4–7 mm long, beak teeth ≤ 0.5 mm long. Known in Wisconsin from a single 1965 Iowa County collection.

78. *Carex hystericina* Muhlenberg ex Willdenow. Cespitose; basal leaf sheaths reddish to purplish brown, fibrillose; culm angles scabrous beneath the inflorescence; pistillate spikes oblong or narrowly oblong; perigynia slightly inflated, ≤ 2 mm wide, 13–20-veined, the veins mostly fusing by the middle of the beak, teeth of the perigynium beak < 1 mm long (pp. 148–149).

79. *Carex lurida* Wahlenberg. This plant suggests a robust *C. hystericina* [78] with perigynium veins fewer (typically 8–12) and distinct nearly to the tip of the beak. Culm angles scabrous beneath the inflorescence; pistillate spikes > 1.5 cm thick; perigynia strongly inflated, lustrous, ≤ 10 mm long, > 2.5 mm wide. Floodplains of the Upper Wisconsin, Black, and LaCrosse rivers.

80. *Carex oligosperma* Michaux. Colonial, loosely cespitose or shoots arising singly; rhizomes long; leaf blades wiry; staminate spike solitary; perigynia 3–15 per spike, inflated, lustrous (pp. 150–151).

81. *Carex tuckermanii* Dewey. Cespitose; basal leaf sheaths red to purple; pistillate spikes ascending to drooping, oblong to thickly cylindrical; perigynia inflated, short-beaked, lustrous; achenes notched on 1 side (pp. 152–153).

82. *Carex retrorsa* Schweinitz. Similar to *C. hystericina* [78] and *C. lurida* [79], but culms smooth beneath the inflorescence; pistillate scales inconspicuous, neither awned nor scabrous; perigynia spreading to reflexed, inflated, 6–10 mm long. Floodplains, wet forests, and other wetlands throughout Wisconsin.

83. *Carex vesicaria* Linnaeus. Often mistaken for *C. utriculata* [84], but plants cespitose; rhizomes short; plant bases and basal leaf sheaths not spongy; ligules longer than wide (associate this with the V in *C. vesicaria*); culms typically scabrous on the angles beneath the inflorescence (this appears to be the least reliable of the characters). Sunny swales in alluvial wetlands throughout the state.

84. *Carex utriculata* W. Boott [= *C. rostrata* var. *utriculata*]. Colonial; rhizomes elongate; plant bases and basal leaf sheaths spongy; leaf blades septate-nodulose, ligules at most as long as wide; culms usually smooth or somewhat scabrous on the angles beneath the inflorescence; pistillate spikes ascending or nodding, oblong to cylindrical; perigynia inflated, strongly veined, short-beaked (pp. 154–155).

85. *Carex rostrata* Stokes. Similar to *C. utriculata* [84], but leaves ≤ 5 mm wide, U-shaped in cross-section, papillose on the upper (adaxial) surface. Boggy habitats, northern Wisconsin.

Section *Paludosae* G. Don

This is a heterogeneous section. Within our flora, only the common lake sedge (*Carex lacustris* [86]) possesses the style character and general appearance of the bladder/bottlebrush sedge group. The other three species [87–89], formerly set apart as section *Hirtae* Mackenzie, are slender, long-rhizomatous, with basal leaf sheaths red (ladder-fibrillose in all but *C. houghtoniana* [87]) and perigynia pubescent.

86. *Carex lacustris* Willdenow. Plants colonial; basal leaf sheaths spongy, red, ladder-fibrillose; leaf blades W-shaped in cross-section, 8–20 mm broad; perigynia very short-beaked, bidentate or some bearing a tiny third beak tooth (pp. 156–157).

87. *Carex houghtoniana* Torrey ex Dewey. Similar to *C. pellita* [88], but plants generally shorter, stouter, basal leaf sheaths not ladder-fibrillose. Perigynia 4.5–6.5 mm long, not densely pubescent, veins evident. Dry to moist sandy and rocky soils, predominantly in the northern third of the state.

88. *Carex pellita* Willdenow [= C. *lanuginosa* auct. non Michaux]. Leaf blades flat or M-shaped in cross-section, the broadest ≥ 2.5 mm; perigynia 2.5–5 mm long, densely pubescent (pp. 158–159).

89. *Carex lasiocarpa* Ehrhart. Similar to *C. pellita* [88] and often mistaken for it, but leaf blades involute, wiry, ≤ 2 mm wide; perigynia 3–4.5 mm long. Standing water or sphagnum bogs, where it forms floating mats, throughout most of Wisconsin except the Driftless Area. Wisconsin plants are *C. lasiocarpa* ssp. *americana* (Fernald) Hultén.

Section *Carex*

*Plants typically colonial; rhizomes elongate. **Vegetative culms** prominent. **Perigynia** long-beaked with prominent beak teeth.*—Leaf sheaths provide some of the most distinctive characters for identification in this section. This is fortunate, as the tall vegetative culms generally overtop and outnumber the fertile culms.

90. *Carex trichocarpa* Muhlenberg ex Willdenow. Leaf sheaths and blades glabrous; inner band of the leaf sheaths stained dark reddish or purple at the summit; perigynia pubescent, the beak prominent, long-toothed (pp. 160–163).

91. *Carex hirta* Linnaeus. Leaf sheaths and inflorescence scales pubescent; perigynia 5–8 mm long, pubescent. European, known in Wisconsin from two highly disturbed sites in Wood and Grant counties, where it may no longer persist.

92. *Carex atherodes* Sprengel. Leaf sheaths pubescent, rarely glabrous; leaf blades finely papillose on the underside (abaxial surface), at least the lower blades sparsely pubescent toward the base; perigynia 7–12 mm long, glabrous or slightly pubescent on the beak only. Sedge meadows, marshes, willow swamps, tamarack bogs, often in standing water, mostly in the southeastern quarter of the state, scattered populations in the north and west (pp. 160–163).

93. *Carex laeviconica* Dewey. Similar to *C. atherodes* [92], but leaf sheaths glabrous, the summit of the inner band strongly veined; leaf blades glabrous, not papillose on the underside (abaxial surface); perigynia 5–8.5 mm long, glabrous. Floodplains of the Mississippi and Lower Wisconsin rivers, often on banks, levees, and drainage spoils (pp. 160–163).

Section *Rostrales* Meinshausen [= *Folliculatae* Mackenzie]

Plants cespitose; rhizomes elongate; bases brown, not reddish or purplish. Staminate spike solitary, the base lower than or roughly equaling the apex of the uppermost pistillate spike. Pistillate spikes 2–6, approximately as long as thick. Perigynia ≤ 20 per spike, divergent or the lowermost reflexed, somewhat inflated, narrow, tapering continuously to the apex, generally 4–7 times as long as wide, beakless, subtly bidentate.—Species in this section superficially resemble *Carex intumescens* [96], but they differ both in their relatively drab bases and in their narrower perigynia.

94. *Carex folliculata* Linnaeus. An elegant plant, distinctive by the tufts of lanceolate perigynia that make up the pistillate spikes. Leaf blades 6–20 mm wide; inner band of the bract sheaths prolonged at the summit, often forming a short, membranous appendage; perigynia 10–15 mm long, 2–3 mm wide. Locally abundant in wet forests, boggy thickets, and sandy wetlands in west-central Wisconsin.

95. *Carex michauxiana* Boeckeler. Similar to *C. folliculata* [94], but widest leaf blades < 5 mm wide; bract sheaths not prolonged, concave at the apex; perigynia 9–12 mm long, 1–2 mm thick. Locally abundant in bogs, bottomland forests, and wet acidic soils in the Bayfield area.

Section *Lupulinae* Tuckerman ex J. Carey

These are among the loveliest species in the North American flora, gracing wet forests throughout the state. Species in the section are easily recognized by having *perigynia greatly inflated, strongly ribbed, 1–2 cm long, beak teeth short, firm, distinct.*

96. *Carex intumescens* **Rudge.** Cespitose with evergreen leaves; perigynia generally ≤ 12 per pistillate spike, mostly ascending, lustrous; styles straight (pp. 164–167).

97. *Carex grayi* **J. Carey.** Similar to *C. intumescens* [96], but pistillate spikes globose; perigynia radiating outward in all directions, 12–20 mm long, dull, sometimes pubescent (pp. 164–167).

98. *Carex lupulina* **Muhlenberg ex Willdenow.** Suggesting a very robust *C. intumescens* [96]: pistillate spikes longer than thick with up to 80 perigynia; perigynia 11–19 mm long, shiny, glabrous; achenes typically ≤ 2.6 mm wide, angles rounded to acute (not knobbed); styles abruptly bent (pp. 164–167).

99. *Carex lupuliformis* **Sartwell ex Dewey.** Similar to *C. lupulina* [98] but more robust; achenes typically ≥ 2.4 mm wide, angles knobbed. Rare in wet forests in southeastern and south-central Wisconsin along the Baraboo and Wisconsin rivers.

Carex Subgenus *Vignea*

Bracts often setaceous, rarely foliose. Cladoprophylls lacking. Spikes sessile, branching in some species but more often simple, typically bisexual and all similar to one another on a given plant. Perigynia biconvex or planocon-vex (subterete in Carex disperma *[123]). Achenes lenticular. Stigmas 2 in our flora except in aberrant individuals, though regularly 3 in some members of the subgenus not present in the Midwest.*—Subgenus *Vignea* is much smaller than subgenus *Carex,* with between 400 and 500 species distributed primarily in the New World and Eurasia. It is morphologically distinctive, most easily recognized in the field by the bisexual, sessile (unstalked) spikes, which are typically all similar to one another on a given plant. This dif-fers from the condition of most members of subgenus *Carex,* in which the lower spikes are predominantly or wholly pistillate, the upper spikes are largely or altogether staminate, and at least some spikes on each plant are typically stalked (pedunculate). Lower spikes of some subgenus *Vignea* spe-cies (viz., members of sections *Vulpinae, Multiflorae, Heleoglochin,* and some *Phaestoglochin*) are branched. Perigynia of subgenus *Vignea* also have a distinctive 2-faced look, planoconvex or biconvex with, usually, 2 distinct margins; a few, such as *C. disperma* [123] and a few members of section *Glareosae,* are nearly round in cross-section (terete) or have rounded or indistinct margins. Stigmas in most species are forked (bifid), and achenes are almost always lens-shaped. Both of these characteristics are related to the fact that the ovaries are 2-carpellate. (The 3-carpellate ovaries of most subgenus *Carex* species produce achenes that are triangular or round in cross-section.)

Section *Divisae* H. Christ ex Kükenthal

This section of strongly rhizomatous, unisexual plants is more diverse in western North America, where identifications can be difficult, especially of staminate specimens.

100. *Carex praegracilis* W. Boott. This plant is not native in most of eastern North America but has spread rapidly from the west in recent decades, especially along expressways, where it is tolerant of road salt. Plants aggressively and strongly clonal; shoots arising singly from stout, scaly rhizomes; basal leaf sheaths dark, bladeless; inflorescence unisexual, generally ≥ 1 cm long; perigynia 2.5–3.5 mm long, veinless or obscurely veined. Pistillate individuals are reminiscent of *C. sartwellii* [102] or *C. siccata* [103]. However, the hyaline inner band of the leaf sheath on *C. praegracilis* distinguishes it from *C. sartwellii*, and perigynia in *C. siccata* are generally ≥ 4 mm long. Moreover, inflorescences of *C. sartwellii* and *C. siccata* are bisexual, although *C. siccata* will often have some unisexual spikes. Occasional along highways and in barrens throughout the state.

Section *Physoglochin* Dumortier [= *Dioicae* (Tuckerman) Pax]

A northern section with only one species in the United States.

101. *Carex gynocrates* Wormskjöld ex Drejer [= *C. dioica* Linnaeus ssp. *gynocrates* (Wormskjöld ex Drejer) Hultén]. Finely rhizomatous, culms arising singly or in dense tufts; leaf blades very fine, wiry; inflorescence unisexual, less often androgynous; perigynia divergent at maturity, plump (pp. 170–171).

Section *Holarrhenae* (Döll) Pax [= *Distichae* Rouy]

This section as circumscribed in *FNA* contains a single New World species. The section resembles sections *Divisae* and *Ammoglochin* but is easily distinguished from them by the green-veined inner band of its leaf sheaths.

102. *Carex sartwellii* Dewey. Plants strongly colonial; rhizomes elongate; vegetative culms tall, prominent, often more common than the fertile culms; inner band of the leaf sheaths green, strongly veined, hyaline at the summit; inflorescence densely flowered, erect to arching, ≥ 2.5 cm long, all but the lowest spikes overlapping, often indistinct; perigynia 2.5–4 mm long, veined on both faces (pp. 172–173, with *C. siccata* [103]).

Section *Ammoglochin* Dumortier

This predominantly Old World section has a single New World species that is widespread across North America. Its Eurasian relative, *Carex arenaria,* is introduced at a few sites along the eastern seaboard but easily distinguished by its broadly winged perigynia.

103. *Carex siccata* Dewey [= *C. foenea,* misapplied]. Plants colonial, shoots arising singly; rhizomes elongate, scaly; inner band of the leaf sheaths hyaline, at least at the summit; spikes pistillate, staminate, or androgynous, the lower often separate but all typically overlapping, the upper often congested, indistinct from one another; perigynia typically veined on both faces, occasionally veinless on the inner face (pp. 172–173).

Section *Multiflorae* (J. Carey) Kükenthal

Plants cespitose; bases fibrous, brown or pale. Inner band of the leaf sheaths hyaline, corrugated (in Wisconsin species). Culms narrow, firm. Inflorescence compound, cylindrical, densely flowered, stiff. Bracts setaceous. Spikes androgynous, at least the lowest branched. Perigynia planoconvex, weakly or inconspicuously spongy at the base. Primarily in wetlands, throughout the state. — Recognize this section in the field by the corrugated inner band of the leaf sheaths; firm, narrow culms; and densely flowered, straight, compound inflorescence. The section is morphologically more variable in western North America.

104. *Carex vulpinoidea* Michaux. Leaf blades longer than the culms; inflorescence 7–10 cm long with numerous setaceous bracts; perigynia 2–3 mm long, beak ≥ one-third the total perigynium length (pp. 174–175).

105. *Carex annectens* (E. P. Bicknell) E. P. Bicknell [= *C. brachyglossa* Mackenzie]. Nearly as common as *C. vulpinoidea* [104] within its range and often mistaken for it, *C. annectens* differs most obviously in having leaves shorter than the flowering culms and perigynia yellowing at maturity, abruptly narrowed to a short beak. Wisconsin plants are traditionally recognized as *C. annectens* var. *xanthocarpa* (Kükenthal) Wiegand, but this variety appears to intergrade with the typical variety. Leaf blades shorter than the flowering culms; inflorescence 4–7 cm long, 3–5 bracts prominent, several less conspicuous; perigynia yellow at maturity, 2–3 mm long, abruptly short-beaked, beaks ≤ one-third the total perigynium length. Mesic to wet prairies, meadows, and other open, typically sandy areas, primarily in the western half of Wisconsin (pp. 174–175, with *C. vulpinoidea* [104]).

Section *Heleoglochin* Dumortier [= *Paniculatae* G. Don]

Plants densely cespitose; bases brown. Culms narrowing toward the tip, typically arching at maturity. Inner band of the leaf sheaths smooth, pigmented toward the summit. Leaf blades ≤ 3 mm wide in Wisconsin species. Spikes androgynous, the lower branched. Perigynia planoconvex to biconvex, darkening at maturity, mostly ≤ 3 mm long; beak short-triangular, scabrous on the margin, bidentate. Wetlands, primarily peaty, mostly in the eastern half of the state. — As in section *Vulpinae*, species in section *Heleoglochin* are recognized in the field largely by leaf sheath characters. Pigmentation on the inner surface of the leaf sheaths and small, often dark perigynia distinguish this section from the similar sections *Vulpinae* and *Multiflorae*.

106. Carex diandra Schrank. Inner band of the leaf sheaths whitish-hyaline, conspicuously red-dotted at least near the summit; leaf blades 1–2.5 mm wide; ligules ≥ 8 mm long; mature perigynia dark, biconvex, glossy (pp. 176–177).

107. Carex prairea Dewey ex Wood. This species is similar to *C. diandra* [106], but it occupies more calcareous habitats, and the inner band of its leaf sheaths is copper-colored at the summit. Leaf blades 2–3 mm wide; ligules typically < 6 (occasionally 9) mm long; mature perigynia planoconvex, brown, dull (pp. 176–177, with *C. diandra* [106]).

Section *Vulpinae* (Heuffel) H. Christ

Plants cespitose; bases generally pale. Inner band of the leaf sheaths hyaline, in other regards various: corrugated or smooth, thickened or fragile at the summit, sparsely purple-dotted or lacking pigmentation, and combinations of the above. Culms thick, spongy, weak, the angles narrowly winged, scabrous. Inflorescence longer than wide in eastern North American taxa, ovate to cylindrical. Bracts (in eastern North American taxa) setaceous. Spikes densely flowered, the lower branched, mostly or all androgynous (the terminal always androgynous). Perigynia planoconvex, bases spongy (not spongy in Carex alopecoidea *[111]). Wetlands.* — Recognize this section most readily in the field by the thick, spongy culms, branched lower spikes, and spongy perigynium bases (except *C. alopecoidea* [111]). Considered across its entire geographic range, this section grades morphologically into section *Multiflorae*, but in our flora the sections are easily distinguished. Occasional individuals of section *Phaestoglochin* with compound lower spikes (especially *C. sparganioides* [113], *C. cephaloidea* [114], and *C. gravida* [115]) key to this section, but they do not have the thick, spongy, weak culms of section *Vulpinae*.

108. Carex stipata Muhlenberg ex Willdenow. Basal leaf sheaths of the previous year absent; inner band of leaf sheaths corrugated, hyaline, firm at the summit, unpigmented; ligules acute, ≤ 10 mm long; perigynia roughly triangular in outline, 4–6 mm long, base spongy (pp. 178–179). Wisconsin plants are *C. stipata* var. *stipata*.

109. Carex crus-corvi Shuttleworth ex G. Kunze. A long-branched inflorescence, elongate, parallel-sided perigynium beaks, and abrupt, strongly expanded spongy perigynium bases distinguish this species. Basal leaf sheaths of the previous year persistent; inner band of leaf sheaths smooth, red-spotted, fragile; ligule approximately 2 mm long, apex rounded; inflorescence long-branched; perigynia ≤ 8 mm long, base spongy, abruptly thickened, distinct from the remainder of the perigynium body, approximately torus-shaped in many individuals, perigynium beak 2–3 times as long as the body with nearly parallel sides, slender. State endangered species, known from a few populations in Milwaukee and Waukesha counties.

110. *Carex laevivaginata* (Kükenthal) Mackenzie. This attractive species is similar in appearance to *C. stipata* [108] but is distinguished, often even at a distance of a meter or so, by the more robust spikes and perigynium beaks, which darken at maturity. Moreover, the inner band of the leaf sheath in *C. laevivaginata* is smooth, not corrugated, and thickened at the summit. Basal leaf sheaths of the previous year absent; inner band of the leaf sheaths smooth, thickened and yellow at the summit, otherwise hyaline; ligules ≤ 5 mm long, apex acute; perigynia ≤ 6 mm long, both faces strongly nerved, base spongy, apex browning at maturity. State endangered species known in Wisconsin only from a grazed sedge meadow in Dane County, the bottom of a ravine in Iowa County, and the margins of a gravel-bottom river in Monroe County.

111. *Carex alopecoidea* Tuckerman. Superficially similar to *C. stipata* [108], but spikes more compact, leaf sheaths smooth, and perigynia shorter, lacking spongy bases. Basal sheaths of previous years persisting as fibers; inner band of the leaf sheaths hyaline, smooth, not thickened at the summit, often red- or purple-spotted; ligules approximately 5 mm long, apex rounded; perigynia ≤ 4 mm long, inner face veinless, base not swollen. Floodplains, primarily in the southern two tiers of Wisconsin counties and counties adjoining Lake Winnebago, with outliers in St. Croix, Bayfield, Brown, Wood, and Portage counties.

Section *Chordorrhizae* (Heuffel) Meinshausen

A section of only two species. Wisconsin's species is the more widespread of the two, circumpolar, distributed across northern North America and Eurasia. The second species, *Carex pseudocuraica*, is strictly Eurasian.

112. *Carex chordorrhiza* Ehrhart ex Linnaeus. Plants prominently stoloniferous, shoots arising singly; rhizomes short; inflorescence capitate; perigynia dark brown, glossy, strongly veined, abruptly short-beaked (pp. 180–181).

Section *Phaestoglochin* Dumortier [= *Bracteosae* Pax]

*Plants cespitose; rhizomes short or inconspicuous; bases pale to brown, occasionally reddish. **Inner band of the leaf sheaths** hyaline, corrugated or smooth.*

*Spikes all or mostly androgynous, simple in most taxa, the lower branched in some species [113–116]. **Perigynia** mostly planoconvex, beaks typically bidentate.*—This is the most morphologically heterogeneous and probably the most unnatural section in subgenus *Vignea*, with two distinctive groupings in our flora: species with leaf sheaths loose, the backs whitened between the veins with conspicuous cross-veins, inflorescence simple or compound [113–116]; and species with leaf sheaths tight, green or whitened between the veins, lacking conspicuous cross-veins, inflorescence simple [117–122, though the inflorescence is sometimes compound in 119]. The first group appears to be related to sections *Multiflorae* and *Vulpinae*, but relationships within the section are not known.

113. *Carex sparganioides* Muhlenberg ex Willdenow. The elongate, often arching inflorescence of this species, with spikes distantly spaced on the inflorescence axis, suggests a robust *C. rosea* [117] or a diminutive bur-reed (*Sparganium* sp., hence the specific epithet). Occasional individuals nearly overlap *C. cephaloidea* [114] in inflorescence length, but the lowest inflorescence internode is twice as long as the lowest spike in *C. sparganioides*, rarely as long in *C. cephaloidea* [114]. Lower leaf sheaths loose, inner band frequently corrugated, back whitened between the veins with prominent cross-veins; leaf blades 5–10 mm wide; lowest spikes often branched; perigynia green, margins narrow. Mesic forests of southern Wisconsin.

114. *Carex cephaloidea* (Dewey) Dewey. Lower leaf sheaths loose, inner band frequently corrugated, back whitened between the veins with prominent cross-veins; leaf blades 4–8 mm wide; culms 2.5–4 mm thick at the base, ≤ 1 mm thick below the inflorescence; inflorescence dense, stiff; spikes overlapping; pistillate scales concealing ≤ half of the perigynium, apex acute to acuminate, sometimes short-awned; perigynia green, narrowly to indistinctly winged (pp. 182–183).

115. *Carex gravida* L. H. Bailey. Very similar to *C. cephaloidea* [114], distinguished by its pistillate scales, which are acuminate or awned and typically conceal ≥ half of the perigynium body, and perigynia, which often become dark brown and swollen (gravid) at maturity. The inflorescence in this species is occasionally compound. Dry prairies, mostly along disturbed edges, and other open disturbed areas in the southern half of Wisconsin, most likely adventive from the west.

116. *Carex aggregata* Mackenzie. Known from a single weedy railroad right-of-way in Lafayette County; very similar to *C. gravida* [115]. Identifying this species requires material with intact leaf sheaths. Inner band of the leaf sheaths thickened toward the summit, white to yellow or brown, sometimes red-dotted or corrugated, back often white-spotted; widest leaf blades 3.5–5 mm wide; inflorescence densely flowered, longer than wide; spikes mostly or all androgynous; perigynia 3.5–4.5 mm long, base spongy-thickened.

117. *Carex rosea* Schkuhr ex Willdenow [= *C. convoluta* Mackenzie]. Widest leaf blades 1.8–2.5 mm wide; culms leaning or erect throughout the growing season, > 1.5 mm wide at the base; perigynia 2.5–4 mm long, base spongy, the spongy portion ≤ one-fifth the total perigynium length; stigmas thick, tightly coiled (pp. 184–185).

118. *Carex radiata* (Wahlenberg) Small [= *C. rosea* Schkuhr ex Willdenow, misapplied in many floras]. This species and *C. rosea* [117] are frequently confused with one another. Look for finer, straighter stigmas on *C. radiata*, usually bent downward at the base and trailing along the margins or face of the perigynia, sometimes loosely curled. Leaves and culms are narrower in *C. radiata* and tend to sprawl by midsummer, while those of *C. rosea* merely lean or remain erect into late summer. Many taxonomic treatments follow Mackenzie in misapplying the name *C. rosea* to this species and using the name *C. convoluta* for the true *C. rosea*. When you see the name *C. convoluta* in a taxonomic treatment, it refers to *C. rosea;* when the name *C. rosea* appears in the same treatment, it refers to *C. radiata*. Widest leaf blades 1.3–1.9 mm wide; culms ≤ 1.5 mm wide at the base; spikes distant from one another; perigynia 2.5–4 mm long, base spongy, the spongy portion ≤ three-tenths of the total perigynium length; stigmas wavy or straight, occasionally loosely curled, not tightly coiled. Upland to occasional low forests throughout Wisconsin, often with *C. rosea* [117], usually in moister soils (pp. 184–185, with *C. rosea*).

119. *Carex muehlenbergii* Schkuhr ex Willdenow. Like a robust *C. cephalophora* [121] with large perigynia and spikes, forming tall clumps. Plants densely cespitose; culms often to nearly 1 m tall; inner band of the leaf sheaths often corrugated, yellowish and thickened at the summit; inflorescence coarse, robust, distinctly longer than wide; spikes densely aggregated; pistillate scales ≥ two-thirds as long as the perigynia; perigynia 3–4 mm long, strongly veined on the back. Both *C. muehlenbergii* var. *muehlenbergii* and *C. muehlenbergii* var. *enervis* Boott occur in Wisconsin. Distinguish the two by pistillate scale length (2.5–3.5 mm in the typical variety, 2–3.5 mm in var. *enervis*) and perigynium length (3–4 mm in the typical variety, 2.7–3 mm in var. *enervis*). Venation on the inner face of the perigynium, usually present in the typical variety but always lacking in var. *enervis*, is a less reliable character. Prairies, sand barrens, and other dry, frequently disturbed sites, primarily in the southern half of the state (pp. 186–187, with *C. cephalophora*).

120. *Carex spicata* Hudson. This Eurasian species is known in Wisconsin from a small handful of lawns and pastures in the southern half of the state. The reddish coloration at the base of the plant is difficult to see but exceptional in the subgenus. Basal leaf sheaths tinged red or purple, often sparsely; ligules 4–8 mm long; inflorescence loose, the lowest spikes separate or all spikes overlapping; perigynia glossy black at maturity, 4–5.5 mm long.

121. *Carex cephalophora* Muhlenberg ex Willdenow. Leaf sheaths tight, smooth, the summit of the inner band subtly thickened; leaf blades 2–5 mm wide; inflorescence capitate, 1–2 times as long as wide, bracts typically conspicuous, short, setaceous; pistillate scales acuminate to short-awned at the apex, 1–2 mm long; perigynia 2.5–3 mm long, widest near the middle, beak teeth 0.3–0.5 mm (pp. 186–187).

122. *Carex leavenworthii* Dewey. Similar to *C. cephalophora* [121], but pistillate scales acute to short-toothed at the apex, 1.5–2.5 mm long; perigynia widest below the middle; beak teeth 0.1–0.3 mm. Rare, known in south-central Wisconsin from very few collections in lawns and gardens.

Section *Dispermae* Ohwi

This section contains a single species that is morphologically distinctive and distributed across northern North America and Eurasia.

123. *Carex disperma* Dewey. Plants fine, shoots arising singly or in small bunches from pale, slender rhizomes; spikes few-flowered, androgynous; perigynia spreading, darkening at maturity, plump, biconvex to subterete in cross-section (p. 188).

Section *Glareosae* G. Don [= *Heleonastes* Kunth, *Canascentes* Fries ex Kükenthal]

Plants cespitose; rhizomes various. Inner band of the leaf sheaths hyaline. Spikes distinct, mostly nonoverlapping (except Carex arcta *[126], which has spikes overlapping, the upper indistinct from one another), mostly or all gynecandrous, lateral spikes sometimes pistillate. Perigynia ascending to spreading; margins rounded in most species, smooth or finely serrate, often finely papillose. Cool, wet soils, often peaty, more common northward.*—This section is widespread in Eurasia, Canada, and the northern United States. It is easily recognized by the gynecandrous spikes with perigynia not divergent (as in section *Stellulatae*) or, for the most part, acutely margined (as in sections *Stellulatae, Deweyanae,* and *Ovales*). The oddball in the section is *C. arcta* [126], which is reminiscent of a bushy-headed member of section *Phaestoglochin* but with gynecandrous spikes. Get a feel for the section by comparing the illustrations of *C. trisperma* [125], *C. canescens* [127], and *C. brunnescens*

[128] with those of sections *Stellulatae* (*C. interior* [133] and *C. echinata* [134]) and *Deweyanae* (*C. deweyana* [129] and *C. bromoides* [130]).

124. *Carex tenuiflora* Wahlenberg. This rare species of calcareous swamps is the only *Vignea* species in the Wisconsin flora to have gynecandrous spikes and essentially beakless perigynia. Plants loosely cespitose; rhizomes elongate, slender; leaf blades 0.5–2 mm wide; inflorescence as long as wide or slightly longer; spikes 2–4, overlapping; pistillate scales 3-veined, translucent, whitish except for the green keel; perigynia gray-green, 3–3.5 mm long, weakly veined, beakless or nearly so. Northern Wisconsin and Cedarburg Bog (Ozaukee County).

125. *Carex trisperma* Dewey. Plants loosely cespitose; rhizomes elongate, slender; culms arching, slender; inflorescence very open, lowest bract equaling or overtopping the tip of the inflorescence; spikes 1–3, widely separated; perigynia 1–5 per spike (p. 189). Wisconsin's plants are probably all *C. trisperma* var. *trisperma*, but *C. trisperma* var. *billingsii* O. W. Knight may also be present in the state.

126. *Carex arcta* Boott. This species is recognized among other members of the section by its many-flowered spikes overlapping in a dense inflorescence, the upper spikes often indistinct; and perigynia that are widest at the base, the beaks finely serrate along the margin, suggesting the perigynia of section *Stellulatae*. Plants densely cespitose; rhizomes short; inner band of the leaf sheaths hyaline, purple-dotted; leaf blades pale or grayish; inflorescence densely flowered, the upper spikes frequently indistinguishable; perigynia ascending to spreading, green to brown, 2–3.5 mm long, widest near the base, beak approximately half as long as the perigynium body, finely serrate on the margin. Uncommon in northern wet forests, often in floodplains.

127. *Carex canescens* Linnaeus. Plants cespitose; rhizomes short; leaf blades pale green to grayish, 2–4 mm wide; spikes ovoid, distinctly longer than wide; perigynia typically 10–20 per spike, 2–3 mm long, beak margin smooth or serrulate at the base (pp. 190–191).

128. *Carex brunnescens* (Persoon) Poiret. Plants cespitose; rhizomes short; leaf blades green, 1–2.5 mm wide; spikes approximately as long as wide to slightly longer; perigynia 5–10 per spike, 2–2.5 mm long, beak margin very finely serrate (pp. 192–193).

Section *Deweyanae* (Tuckerman ex Mackenzie) Mackenzie

Plants cespitose; rhizomes mostly short; bases brown. **Inflorescence** *slender, open, at least the lowest spike(s) distinct; bracts setaceous.* **Spikes** *mostly gynecandrous, lateral spikes sometimes pistillate, mixed, or (rarely) staminate.* **Perigynia** *appressed to ascending, ovate to lanceolate, planoconvex, slender;*

base spongy; beak distinct, margins serrate, apex bidentate. **Achenes** *mostly filling the perigynium body. Cool, dry-mesic to, more often, mesic or wet soils; usually shaded.* — Several taxa in western North America can be difficult to distinguish from one another, but the two midwestern species are among our most common and easiest to identify. The section is distinguished from the other gynecandrous *Vignea* sections by the narrow, appressed to ascending, acute-margined to very narrowly winged perigynia. Perigynia of section *Glareosae* are mostly plumper with rounded margins, those of section *Stellulatae* broadest at the base and divergent to reflexed at maturity, and those of section *Ovales* prominently winged and, with a few exceptions, wider than those of section *Deweyanae*.

129. *Carex deweyana* Schweinitz. Plants densely cespitose; rhizomes short; widest leaf blades 2.5–4 mm wide; upper spikes overlapping, lowest spikes separated by 1–3 cm; perigynia 4–5 mm long, 3–4 times as long as wide, veinless to few-veined on the inner face, veinless on the back. Wisconsin's plants are *C. deweyana* var. *deweyana* (pp. 194–195).

130. *Carex bromoides* Schkuhr ex Willdenow. This plant forms dense clumps in bottomlands and forested seeps throughout much of the state, especially northward, where it is easily recognized by its sprawling, wispy foliage and lanceolate perigynia. Plants densely cespitose; rhizomes sometimes elongate, with internodes to 2 cm; culms weak, 20–90 cm tall; widest leaf blades 1.5–3 mm wide; inflorescence slender; spikes gynecandrous, less often pistillate, mostly or all overlapping, the lowest separated by 3–20 mm; perigynia lanceolate, 4–7 mm long, > 4 times as long as wide, strongly to weakly veined on the inner face, usually strongly veined on the back. Wisconsin's plants are *C. bromoides* ssp. *bromoides* (pp. 194–195, with *C. deweyana*).

Section *Stellulatae* Kunth

Plants cespitose; rhizomes short; bases brown, not fibrous. **Inflorescence** *mostly open, spikes readily distinguished from each other, the lowest in our more common species not overlapping; bracts inconspicuous or lacking.* **Spikes** *2–10 (solitary in* Carex exilis *[131]), gynecandrous (unisexual in* C. sterilis *[132]).* **Perigynia** *spreading to reflexed, typically planoconvex, widest at the base, generally chestnut brown to dark brown or even blackish at maturity; margins acute; base spongy; beak generally bidentate, margins finely serrate.* **Achenes** *much smaller than the perigynia. Acidic and calcareous wetlands throughout the state.* — This is an extremely difficult section in areas where the species are more numerous, but Wisconsin's four species are not difficult to identify.

131. *Carex exilis* Dewey. This species is unique in the section both by the fact that it is unispicate and by its involute leaves (all other species in the section have leaves flat or V-shaped in cross-section). Leaf blades involute, 0.5–1.5 mm wide, the widest on a given plant often > 1 mm wide; spikes typically solitary, gynecandrous, 0.5–4 cm long, occasionally unisexual or accompanied by smaller lateral spikes, then forming a headlike inflorescence; perigynia chestnut to dark brown, 2.5–4.5 mm long, faintly veined on the inner face; anthers 2–3.5 mm long, longer than in any other section *Stellulatae* species in our flora with the occasional exception of *C. sterilis* [132]. Bogs of the Apostle Islands and Door County.

132. *Carex sterilis* Willdenow. One of only three dioecious carices in Wisconsin's flora. Occasional individuals have staminate and pistillate flowers intermixed in the spikes or staminate spikes and pistillate spikes in the same inflorescence. Widest leaf blades 1.5–2.5 mm wide; spikes 3–8, typically unisexual, the upper crowded, at least the lowest separate; perigynia chestnut brown to blackish, 2–4 mm long, veinless to several-veined on the inner face; anthers 1–2.2 mm long. Fens and calcareous wet prairies primarily in central and southeastern Wisconsin, scattered populations in Door County, the Apostle Islands, and northeastern Wisconsin.

133. *Carex interior* L. H. Bailey. Widest leaf blades 1–2.5 mm wide; spikes 2–5 (typically 3), gynecandrous, base of the terminal spike conspicuously clavate; perigynia chestnut to dark brown, usually veinless on the inner face, abruptly beaked, beak < half as long as body (pp. 196–197).

134. *Carex echinata* Murray. Similar in appearance to *C. interior* [133], *C. echinata* is most easily distinguished by the larger number of spikes on most individuals and the perigynium shape, which is more nearly triangular and tapers gradually to the beak. *Carex echinata* also differs in habitat, tending toward more acidic sandy or peaty wetlands than *C. interior.* Widest leaf blades 1–3.5 mm wide; spikes 3–8, typically ≥ 4, gynecandrous, the staminate base of the terminal spike often ≥ twice as long as the pistillate portion; perigynia approximately triangular in outline, tapering gradually to the beak, 3–4 mm long, veinless to several veined on the inner face, beak 1–2 mm long and ≥ half as long as the body. Acidic wetlands, sandy and/or peaty, mostly in the northern half of Wisconsin, scattered in south-central Wisconsin.

Section *Ovales* Kunth

Plants cespitose, usually densely so; rhizomes short; bases brown. Culms both vegetative and fertile, the vegetative culms with solid nodes and typically hollow internodes, more conspicuous in some species than the fertile culms.

Spikes mostly *gynecandrous, lateral spikes sometimes pistillate.* **Perigynia** *flat and scalelike or planoconvex; marginal wings* ≥ *0.2 mm wide at their widest point; bases lacking spongy tissue.* **Achenes** *lenticular, typically relatively small compared to the perigynium body. Dry-mesic to wet soils, full sun to dense shade; most species thrive under moderate disturbance.* — See text under Key J (pp. 49–51).

135. *Carex sychnocephala* J. Carey. The long, foliose bracts and narrow, long-beaked perigynia of this species are highly distinctive. You are not likely to mistake it for anything else in the North American flora. Leaf blades to 3 per culm, 1–3 mm wide; lowest bracts leaflike, 5–20 cm long, > 3 times as long as the inflorescence; inflorescence capitate or the lowest spikes separate; spikes approximately twice as long as wide; perigynia appressed, lanceolate, 5.5–7.5 mm long, 0.5–1 mm wide, beak approximately twice as long as the perigynium body. Sandy or peaty wetlands of central Wisconsin and scattered populations in northeastern Wisconsin, frequently associated with seasonally fluctuating water levels.

136. *Carex muskingumensis* Schweinitz. Plants densely cespitose; rhizomes short, stout, dark; vegetative culms numerous, tall, rounded-triangular to circular in cross-section, leaf blades relatively evenly spaced along the upper half; inner band of the leaf sheaths green-veined to within 3 mm of the summit; spikes 5–12, spindle-shaped, base and apex narrowly tapered; perigynia lanceolate, conspicuously veined on both faces, 6–9 mm long (pp. 198–199).

137. *Carex cristatella* Britton. Similar to both *C. tribuloides* [138] and *C. projecta* [139], distinguished from both by its spherical spikes bearing strongly divergent perigynia, the beaks of which conceal the pistillate scales at maturity. Immature individuals may be confused with *C. bebbii* [154]. That species, however, lacks the expanded sheaths and prominent vegetative culms of *C. cristatella* and relatives. Leaf blades 3–7.5 mm wide; inflorescence erect to arching; spikes 6–15, globose, the upper overlapping, the lower usually separated; perigynia 2.5–4 mm long, conspicuously veined on both faces, spreading to divergent. Alluvial soils, mostly in the southern half and eastern third of Wisconsin.

138. *Carex tribuloides* Wahlenberg. Plants cespitose, often reproducing vegetatively from the previous year's vegetative culms and appearing stoloniferous; vegetative culms numerous, tall, often reclining, leaf blades relatively evenly spaced on the upper half; leaf sheaths loose, expanded near the summit, sharply angled, green-veined on the inner band nearly to the summit except for a hyaline, inverted-triangular region 3–8 mm long; inflorescence erect, rarely arching; spikes mostly overlapping; perigynia ≥ 30 per spike, appressed to ascending, scalelike. Wisconsin's plants are *C. tribuloides* var. *tribuloides* (pp. 200–201).

139. *Carex projecta* Mackenzie. This plant is common in mesic to wet forests of central and northern Wisconsin. Ecologically, it is mostly distinct from *C. tribuloides* [138], which grows almost exclusively on alluvial soils. The two rarely occupy the same site, but when they do grow together, *C. tribuloides* is usually more robust, with wider leaf blades, a stiff inflorescence, and later flowering/fruiting dates, often occupying sites closer to the main channel of a river. *Carex projecta* also tends to have darker foliage, the inner band of the leaf sheaths fragile at the summit, and a hyaline inverted-triangular region 4–20 mm long at the summit. Leaf blades 3–7 mm wide; inflorescence nodding; lower spikes mostly not overlapping; perigynia ≤ 30 per spike, ascending to divergent, 3–4.5 mm long, beaks spreading (pp. 200–201, with *C. tribuloides* [138]).

140. *Carex adusta* Boott. The inflorescence of this species is similar in size and texture to that of *C. muehlenbergii* [119], but with spikes gynecandrous and less densely aggregated. Inner band of the leaf sheaths hyaline; leaf blades 2–3.5 mm wide; inflorescence stiff; lower bracts occasionally foliose, exceeding the tip of the inflorescence; spikes mostly or all overlapping, slightly longer than wide; perigynia ascending, 4–5 mm long, thick, leathery, inner face typically veinless, margins winged above and rounded near the base; achenes relatively large, almost filling the perigynium body. Pine barrens, lakeshores, and other sandy disturbed soils in central and northern Wisconsin.

141. *Carex foenea* Willdenow [= *C. aenea* Fernald]. The graceful, nodding inflorescence of this species is similar to that of *C. straminea* [143], but the pistillate scales of *C. foenea* nearly conceal the perigynia. Inner band of the leaf sheaths hyaline or mottled green and white; leaf blades 2–4 mm wide; inflorescence nodding; spikes all or mostly separate, bases short-clavate; pistillate scales more or less concealing the perigynia; perigynia conspicuously veined on the back, veined or veinless on the inner face, 3.5–5 mm long, tapering to the beak. Rocky soils, primarily in the northern third of Wisconsin. The name *C. foenea* is misapplied to the species *C. siccata* [103] in many manuals; consequently, *C. siccata* and true *C. foenea* may be intermixed in herbarium folders under the name *C. foenea*.

142. *Carex ovalis* Goodenough. The terete, entire-margined beak tips of this species are unique among *Ovales* in the Wisconsin flora but common among *Ovales* of western North America. Leaves 2–3.5 mm wide; inflorescence stiff, erect, the lower spikes usually separate, otherwise all spikes overlapping; pistillate scales equaling the perigynium beak tips, narrower than the perigynium bodies; perigynia conspicuously veined on both faces, 3.5–4.5 mm long, beak cylindrical, unwinged, and entire at the apex for approximately 0.5 mm. Adventive from Europe; the species' only locality in Wisconsin—the yard of the Devil's Island lighthouse—is its farthest inland locality in North America.

143. Carex straminea Willdenow ex Schkuhr. The graceful, arching inflorescence of *C. straminea* is reminiscent of *C. projecta* [139], *C. foenea* [141], and *C. tenera* [157a, 157b], but the green inner band of the leaf sheaths and nearly orbiculate perigynium bodies distinguish *C. straminea* from all three. Inner band of the leaf sheaths green, veined nearly to the summit; leaf blades 1.5–3 mm wide; inflorescence nodding; spikes 3–7, base narrowly clavate, apex rounded; perigynia spreading, body nearly orbiculate, base rounded, 4–5.5 mm long, distinctly veined on both faces. Uncommon in sandy wet meadows, lakeshores, and roadside ditches in west-central Wisconsin.

144. Carex suberecta (Olney) Britton. Stiff inflorescence, conical spike apices, and diamond-shaped perigynia make this species distinct in the occasional southern calcareous fens and meadows that it occupies. Rhizomes sometimes elongate in older plants; inner band of the leaf sheaths distinctly green and veined nearly to the summit; leaf blades 1.5–2.5 mm wide; inflorescence stiff, erect, brown; spikes 2–4 (rarely 5), base rounded or acute, apex acute; perigynia appressed, diamond-shaped with a wedge-shaped base, typically golden brown, 4–5.5 mm long, weakly veined on the inner face, distinctly veined on the back. Rare in fens and calcareous ditches and pastures; Rock, Walworth, and Jefferson counties.

145. Carex cumulata (L. H. Bailey) Mackenzie. Originally described as a variety of *C. straminea* [143], this species actually bears closer resemblance to *C. suberecta* [144] because of its stiff, erect inflorescence and conical spike apices. *Carex cumulata* is readily distinguished from *C. suberecta* by the wider leaf blades; obovate, shorter perigynia; and prominent, bristlelike bracts at the bases of the lower spikes, which are distinctive even in immature individuals. Inner band of the leaf sheaths green and veined nearly to the truncate summit, which extends approximately 0.5 mm beyond the point of connection with the leaf blade; leaf blades 3–6 mm wide; inflorescence erect; spikes overlapping or separated, base rounded, apex conical; perigynia appressed, widest above the middle, inner face noticeably concave, indistinctly veined, the back noticeably convex, 3–4 mm long, 2–3 mm wide. Oligotrophic wet meadows and sphagnous or sandy wet woods in west-central Wisconsin, primarily in Jackson County.

146. *Carex longii* Mackenzie. This species probably does not occur in Wisconsin today. The only collection in the state was made in LaCrosse County in 1958 and was likely an introduced plant. The species grows in Illinois, Indiana, and Michigan. Similar to *C. cumulata* [145] in spike and inflorescence shape, *C. longii* differs in its concave leaf sheath summit, which does not extend past the point of connection with the leaf blade; shaggier-looking spikes, their margins fringed with ascending (not appressed) perigynium beaks; and strong venation on the inner face of the perigynia. In neighboring states also look for *C. albolutescens* (not included in this book as it does not occur in Wisconsin) which has distinctly spreading perigynia beaks and an S-shaped bend at the base of the style (you will need to slice open the perigynium to see this). Vegetative culms often prominent, leaves clustered at apex; inner band of the leaf sheaths green and veined nearly to the summit; leaf blades 2–4.5 mm wide; inflorescence erect; spikes overlapping, base rounded, truncate, or clavate, apex conical; perigynia appressed to ascending, 3–4.5 mm long, 1.5–3 mm wide.

147. *Carex bicknellii* Britton. Inflorescence mostly open, erect to arching, occasionally nodding; perigynia 4.5–7 mm long, 2.5–4 mm wide, distinctly veined on both faces, the epidermis on the inner face translucent; margins brown at maturity (pp. 202–203).

148. *Carex merritt-fernaldii* Mackenzie. Named after Merritt Lyndon Fernald, who described many *Carex* taxa new to science and wrote the magnificent eighth edition of *Gray's Manual of Botany*. This northern species is similar to *C. bicknellii* [147], but the two species barely overlap in geographic range. It is also sometimes confused with *C. brevior* [151], which lacks the papillose leaf sheaths of *C. merritt-fernaldii*. Plants densely cespitose; leaf sheaths papillose (30–40× magnification), inner band hyaline; inflorescence compact or open, occasionally arching; spike base and often the apex rounded; perigynia 3.5–5 mm long, 2.5–3.5 mm wide, at most weakly veined on the inner face, the epidermis translucent, margins yellowish at maturity. Dry sandy soils of central, north-central, and northeastern Wisconsin.

149. *Carex molesta* Mackenzie ex Bright. This common species is usually easily recognized by its terminal cluster of 3–4 nearly spherical spikes and perigynia > 2 mm wide. Similar to *C. brevior* [151] and *C. normalis* [155], it is most easily distinguished from *C. brevior* by the venation on the inner perigynium face and elliptical perigynium body and from *C. normalis* by the broader perigynia and smaller number of more densely clustered spikes. Inner band of the leaf sheaths green, with a hyaline region at the summit; leaf blades 1.5–4 mm wide; inflorescence stiff, congested; spikes 2–4, globose to ellipsoid, rounded at the base and apex; pistillate scales generally extending to the base of the perigynium beak; perigynia 3.5–5 mm long, ≥ 2 mm wide, conspicuously veined on both faces, sometimes veinless on the inner face. Wet prairies, meadows, and wetland edges in the southern half of Wisconsin.

150. *Carex festucacea* Schkuhr ex Willdenow. With its arching or nodding inflorescence, narrow leaves, and often papillose leaf sheaths, *C. festucacea* may be confused with *C. tenera* var. *tenera* [157a]. However, the perigynia in *C. festucacea* are wider, often orbiculate or nearly so, and the terminal spike of *C. festucacea* has a prominent clavate base. Leaf sheaths papillose or smooth, green or whitish between the veins on the back, the inner band hyaline; inflorescence open, arching to nodding; spikes rounded at the apex, base of the terminal spike clavate; perigynia 2.5–4 mm long, the body approximately as wide as long, veinless or inconspicuously veined on the inner face. Wet meadows, mostly in the central part of the state; however, the plant is not widely recognized and may be undercollected in Wisconsin.

151. *Carex brevior* (Dewey) Mackenzie ex Lunell. One of our more common prairie species, *C. brevior* is distinguished from similar species—most notably, *C. bicknellii* [147], *C. molesta* [149], and *C. festucacea* [150]—by the combination of an open, erect or arching inflorescence and wide, typically orbiculate perigynium bodies that are veinless or weakly veined on the inner face. Inner band of the leaf sheaths hyaline; inflorescence open, erect or arching; spikes 4–7, base tapered, apex acute or rounded; pistillate scales generally extending to the middle or tip of the perigynium beak; perigynia 3.5–5 mm long, 2–3 mm wide, the body orbiculate or broadly ovate, veinless or faintly veined on the inner face; achenes orbiculate to broadly ovate, 1.5–2.2 mm long, 1.3–1.8 mm wide. Dry to mesic prairies and meadows in southern Wisconsin (pp. 202–203, with *C. brevior*).

152. *Carex crawfordii* Fernald. Spikes overlapping, longer than wide, usually tapering to an acute apex; perigynia lanceolate or narrowly ovate, typically flat and scalelike, 3.5–4 mm long, approximately 1 mm wide (p. 204).

153. *Carex scoparia* Schkuhr ex Willdenow. Vegetative culms sometimes obvious in the field, leafy at the summit; inflorescence variable, stiff or arching, congested to open; spikes longer than wide, base rounded to acute, apex typically acute; perigynia ascending, lanceolate, 4–7 mm long, 1.2–2 mm wide, conspicuously veined on both faces. Wisconsin plants are *C. scoparia* var. *scoparia* (p. 205).

154. *Carex bebbii* (L. H. Bailey) Olney ex Fernald. Plants densely cespitose; inflorescence congested; lowest bracts bristlelike, often conspicuous; perigynia 2.5–4 mm long, usually veinless on the inner face (p. 206).

155. *Carex normalis* Mackenzie. Stiff inflorescence, broad leaf blades, and whitened areas between the veins on the backs of the leaf sheaths distinguish this species from Wisconsin's other upland *Ovales*. A form with slender, elongate, arching inflorescences (*C. normalis* f. *perlonga*) appears sporadically throughout the range of the species. Vegetative culms sometimes conspicuous, leaves clustered at the apex; leaf sheaths conspicuously whitened on the backs between the veins, the inner band hyaline or green-veined, prolonged at the summit slightly beyond the juncture with the base of the leaf blade; leaf blades 2–6 mm wide; inflorescence erect, occasionally elongate and arching with spikes distant from one another (*C. normalis* f. *perlonga*), 1.5–5 cm long; spikes overlapping or the lowest separate; perigynia spreading, greenish, 2.5–4 mm long, veined on both faces. Dry to mesic woodlands and field/prairie edges throughout the state (p. 207).

156. *Carex tincta* (Fernald) Fernald. This species resembles *C. normalis* [155]—in fact, *C. tincta* was originally described as a variety of *C. normalis* under the older species name *C. mirabilis* Dewey. To a lesser degree it also resembles *C. bebbii* [154] and rare individuals of *C. tenera* var. *tenera* [157a] with congested inflorescences, but it differs in having leaf sheaths green between the veins (as in *C. tenera* var. *tenera* but unlike *C. normalis*) and perigynia typically veined on the inner face (as in *C. normalis* but unlike *C. tenera* var. *tenera*) and becoming brown at maturity (unlike any of the three similar species). Rhizomes of older plants elongate; vegetative culms sometimes evident, leaves clustered at the apex; inner band of the leaf sheaths hyaline, often corrugated; leaf blades 2–4 mm wide; inflorescence stiff, brown, 1.5–3.5 cm long; spikes overlapping; perigynia brown at maturity, 3.5–4.5 mm long, conspicuously veined on both faces. Known only from the airport at Madeline Island (Apostle Islands National Lakeshore); introduced from the East Coast.

157a. *Carex tenera* Dewey var. *tenera*. Leaf sheaths papillose (30× magnification); leaf blades typically ≤ 2.5 mm wide; inflorescence open, nodding from beneath the first spike; perigynium beaks appressed to ascending, exceeding the pistillate scales by ≤ 0.8 mm (pp. 208–209).

157b. *Carex tenera* Dewey var. *echinodes* (Fernald) Wiegand. This variety has nodding inflorescences like those of *C. tenera* var. *tenera* [157a], but leaf characteristics more similar to those of *C. normalis* [155], to which it is more closely related. Despite the fact that the species is ecologically and morphologically distinct, it has not received extensive treatment in any regional flora since Fernald (1950). Plants densely cespitose, often producing far more culms than the typical variety; inner band of the leaf sheaths smooth; leaf blades ranging to wider than those of *C. tenera* var. *tenera*, though the two overlap in leaf width; inflorescence open, nodding from beneath the first spike; perigynium beaks spreading, exceeding the pistillate scales by ≥ 0.7 mm. Floodplain and wet-mesic forests, primarily in the southern half of Wisconsin.

PART 2

Field Guide to Wisconsin Carices

Common Wisconsin Species
of *Carex* Subgenus *Carex*

1. CAREX JAMESII—GRASS SEDGE

section *Phyllostachyae*

After botanist, geologist, and military surgeon Edwin James (1797–1861), who discovered the species.

Grass sedge and its relatives share a highly distinctive inflorescence. The terminal spike (illustrated here) is androgynous with a long, foliose lowermost pistillate scale that resembles the bracts subtending the entire inflorescence in other sections. Flowers May, fruits May to June, most perigynia falling by mid- to late July.

Plants cespitose; rhizomes short; basal leaf sheaths brown, bladeless. **Culms** 5–40 cm tall. **Leaf blades** 1–5 mm wide. **Lateral spikes** 1–5, low, inconspicuous, on spreading or drooping stalks. **Terminal spike** androgynous; *lowest pistillate scale long, foliose,* 1.5–3 mm wide; staminate scales short, rounded, giving the staminate portion of the inflorescence a segmented appearance; staminate flowers 5–12. **Perigynia** 1–4, 4–7 mm long; body globose; beak abrupt, approximately as long as the body.

Habitat and state range. Mostly confined to mesic forests in Dane, Green, and Rock counties, with outliers in LaCrosse, Grant, and Racine counties. The species is not common in the state but is tolerant of disturbance, growing in degraded and high-quality woods alike. Typical associates include basswood, red oak, sugar maple, *Carex hirtifolia, C. hitchcockiana,* puttyroot (*Aplectrum hyemale*), false mermaid (*Floerkia proserpinacoides*).

Similar species. This species is readily confused in Wisconsin only with **Carex backii** (Back's sedge [2]), which is uncommon in dry to mesic forests of Dane County, Juneau County to LaCrosse County, Door County, and far northeastern Wisconsin. Back's sedge has *pistillate scales generally > 2.5 mm wide, surrounding and mostly concealing the perigynia,* and the staminate portion of the inflorescence is relatively inconspicuous.

Unispicate inflorescence of *Carex jamesii,* showing 2 perigynia with orbicular bodies and slender beaks, subtended by the distinctive foliose scales. The slender staminate portion of the spike is evident between the 2 perigynia.

3. *Carex pauciflora*—Few-flowered bog sedge
section *Leucoglochin*
Latin: few-flowered.

The perigynia of this species are reminiscent of the precariously attached ripe fruits of jumpseed (*Polygonum virginianum*) or lop-seed (*Phryma leptostachya*), and in fact they serve a similar purpose: at maturity they spring from the plant when broken off. The perigynia may also catch on fur or clothing, though adaptation for animal dispersal is better developed in some other sedge species. Flowers May to June, fruits June to July, most perigynia falling by mid-August.

Plants *loosely colonial; rhizomes slender; basal leaf sheaths brown, bladeless.* Culms 10–40 cm tall, a secondary shoot commonly emerging from the lowest node. **Leaf blades** few per plant, to 1.5 mm wide. **Spike** *androgynous, solitary;* staminate portion erect, few-flowered, conelike, slender. **Perigynia** 2–6, *reflexed at maturity, lanceolate,* 6–9 (occasionally 10) mm long, distinctly few-veined, *base filled with spongy tissue.*

Habitat and state range. Sphagnum bogs in the northern third of Wisconsin. Associates include *Carex canescens, C. limosa, C. paupercula, C. oligosperma, C. trisperma,* grass pink (*Calopogon tuberosus*), leather-leaf (*Chamaedaphne calyculata*), sundews (*Drosera* spp.), cottongrasses (*Eriophorum* spp.), pitcher-plant (*Sarracenia purpurea*), small cranberry (*Vaccinium oxycoccus*), and other typical northern bog species.

Unispicate inflorescence of *Carex pauciflora.*
The erect cone of staminate flowers is superficially
similar to the 3 reflexed perigynia.

4. CAREX LEPTALEA—BRISTLE-STALKED SEDGE
section *Leptocephalae*
Greek: slender, delicate.

The solitary, few-flowered, androgynous spike and distinctly veined, beakless perigynia of *Carex leptalea* are unmistakable. Flowers May, fruits June to July, perigynia falling in July and August.

Plants cespitose, often forming mats; rhizomes slender, branching; basal leaf sheaths brown. **Culms** 0.1–0.7 m tall, slender. **Leaf blades** generally 2 per culm, flat, to 1.5 mm wide, soft. **Spike** *solitary, androgynous;* staminate flowers few, forming a short cone that emerges from among the perigynia. **Perigynia** 1–10, overlapping, approximately 3 mm long, distinctly and finely veined, tapering to a narrow, nearly stipitate base; *apex blunt or rounded, beakless, dimpled.*

Habitat and state range. Bogs, fens, conifer swamps, occasional sedge meadows, and other peaty wetlands throughout Wisconsin, more common in the eastern half. Common associates include white cedar, tamarack, *Carex brunnescens, C. canescens, C. interior, C. trisperma,* sensitive fern (*Onoclea sensibilis*), marsh fern (*Thelypteris palustris*), marsh marigold (*Caltha palustris*).

Similar species. *Carex leptalea* is geographically wide-ranging and morphologically variable, and as a consequence its taxonomy is not uniformly agreed upon. Unpublished research by Vera Williams and Bruce Ford (University of Manitoba) supports the recognition of three subspecies within *C. leptalea,* of which our plants are the widespread *C. leptalea* ssp. *leptalea.*

Unispicate inflorescence of *Carex leptalea*.
The dimple in the perigynium apex is not visible
in this side view, but the staminate flowers are
obvious, emerging from among the 3 perigynia.

6. CAREX PEDUNCULATA—LONG-STALK SEDGE

section *Clandestinae*

The name of this species derives from the elongate, drooping stalks (peduncles) that bear the pistillate spikes.

This is one of our most widespread northern forest species, recognizable even when not in flower by the elongate, reddish to purple, bladeless leaf sheaths at the bases of the sterile shoots and the leaf blades that die back at the tips. Flowers April, fruits April to May, most perigynia falling by June.

Plants cespitose; rhizomes stout; *basal leaf sheaths of vegetative shoots prominently reddish to purple*, the leaf blades ≤ 1 cm long; basal leaf sheaths of fertile culms reddish or purple with longer leaf blades. **Leaf blades** *overwintering, becoming dark green with brown tips that are separated from the green portion of the leaf blade by a distinct purple line*, generally exceeding the culms, 1.5–4 mm broad. **Culms** weak, drooping at maturity, 8–30 cm long. **Terminal spike** staminate or often androgynous, with a few perigynia at the base. **Lateral spikes** pistillate or androgynous, with a few male flowers at the tip, largely disposed toward the base of the plant, *lower spikes pendulous on slender, elongate stalks (peduncles)*, upper spikes on short stalks. **Pistillate scales** pale to dark brown, midvein extended beyond the margin to form a short awn. **Perigynia** *plump when fresh*, approximately 4.5 mm long, sparsely pubescent above, *base elongate and approximately conical when fresh, shriveled and stalklike when dry*, apex pyramidal with a tiny, abruptly bent beak. The elongate perigynium base in this woodland sedge serves as an elaiosome (oil body) that attracts and is fed upon by ants, which disperse the perigynia.

Habitat and state range. Common in rich, generally calcareous woods throughout Wisconsin; more common northward. Typical associates include *Carex arctata, C. blanda, C. deweyana, C. pensylvanica, C. woodii,* spinulose wood fern (*Dryopteris carthusiana*), shining club moss (*Huperzia lucidula*), princess-pine (*Lycopodium obscurum*), blue cohosh (*Caulophyllum thalictroides*), bunchberry dogwood (*Cornus canadensis*), Dutchman's-breeches (*Dicentra cucullaria*), showy orchis (*Galearis spectabilis*), shin-leaf (*Pyrola chlorantha*), bloodroot (*Sanguinaria canadensis*).

Similar species. *Carex pedunculata* and **C. communis** (**colonial oak sedge** [12]) are two of northern Wisconsin's most common upland forest species. They may be distinguished vegetatively by the *elongate, bladeless, red basal leaf sheaths, narrower leaves,* and *brown leaf tips* of C. *pedunculata* as well as the *more densely cespitose habit* of C. *communis.* See also **C. *richardsonii*** [8].

Plant, androgynous spike, and perigynium of *Carex pedunculata*.
The purple, bladeless basal leaf sheaths, swollen perigynium base,
and slender peduncles are distinctive.

8. *CAREX RICHARDSONII*—RICHARDSON'S SEDGE

section *Clandestinae*

After Sir John Richardson (1827–1865), the Scottish naturalist and explorer of boreal and arctic North America who discovered this species.

In spring and early summer the distinctive brown-purple coloration of the bract sheaths and broad silvery white margins of the staminate scales of this species are easily recognized. Basal rosettes of pale yellow leaves are often the only evidence of this species past the middle of June. Flowers April to May, fruits May to June, perigynia dispersing and culms wilting by the end of June.

Plants mostly loosely cespitose or shoots arising singly; *rhizomes creeping, brown-scaly;* basal leaf sheaths reddish brown, moderately fibrous. **Culms** 8–30 cm tall. **Leaf blades** pale green at first, *becoming yellow to wine red later in the season,* usually shorter than the culms but sometimes extending to > 25 cm long, 1.5–3 mm wide, firm. **Bract sheaths** *maroon.* **Staminate spike** terminal, raised above the pistillate spikes; *anthers transforming the spike into a conspicuous yellow plume in early spring.* **Pistillate spikes** 2–3, clustered near the culm apex. **Perigynia** ≤ 3 mm long, pubescent, tapering to the base; beak abrupt, dark, short, entire or subtly bidentate at the apex.

Habitat and state range. Dry, thin-soiled lime prairies and occasional fens, ranging to dry sand prairies and sand barrens, especially in Burnett and Douglas counties; primarily in the southern third of Wisconsin, less common northward. Typical associates include *Carex eburnea, C. meadii, C. umbellata,* leadplant (*Amorpha canescens*), side-oats grama (*Bouteloua curtipendula*), shooting-star (*Dodecatheon meadia*), puccoon (*Lithospermum canescens*), little bluestem (*Schizachyrium scoparium*), prairie dropseed (*Sporobolus heterolepis*), and, in fens, sedges such as *C. buxbaumii* and *C. tetanica.*

Similar species. This species may be confused with **Carex pensylvanica** (**Pennsylvania sedge [15]**), another early-flowering sedge that often occurs on dry prairies and, like Richardson's sedge, has long-creeping rhizomes and pubescent perigynia. Pennsylvania sedge, however, has *basal leaf sheaths more obviously fibrous, bract sheaths not conspicuously colored,* and the base of the terminal spike typically nestled among upper lateral spikes. There are two other species in section *Clandestinae* in Wisconsin: **C. pedunculata** (**long-stalk sedge [6]**) and **C. concinna** (**beautiful sedge [7]**). Neither has the colored bract sheaths of *C. richardsonii* or grows in dry prairies.

Plant, inflorescence detail, and perigynium of *Carex richardsonii*. Short stature, curled leaves, and early flowering may account for the relative infrequency with which this plant is collected in the field. This illustration displays the staminate spike in its full spring glory, fringed with stamens.

10. *Carex umbellata*—Early oak sedge

section Acrocystis

[= *Carex abdita*, from the Latin for "hidden away," an appropriate name for a sedge with its spikes partly hidden among the leaf bases].

With narrow leaves and fertile spikes partly hidden among the tufted leaf bases, this is one of the more inconspicuous sedges. Look for it in calcareous dry prairies in late April and early May. Flowers April, fruits May, perigynia falling by early to mid-June.

Plants densely cespitose; rhizomes short, often inconspicuous; bases brown, clothed in fibers. **Culms** 3–18 cm tall. **Leaf blades** much longer than the longest culms, 1–2.5 (occasionally 3.5) mm wide. **Terminal spike** staminate, short-stalked. **Lateral spikes** pistillate, 1–5 near the top of the culm, 1–3 near the plant base, compact, < 1 cm long. **Perigynia** 2–3.2 mm long, prominently 2-ribbed, the ribs extending along the margins of the beak, otherwise veinless, pubescent; beaks ≤ 1 mm long.

Habitat and state range. Most common on calcareous dry prairies or dolomite bluffs; occasional on sandstone, sand prairies, and barrens; uncommon in fens and calcareous wetlands. The plant is distributed primarily in southwestern and south-central Wisconsin, a few populations in northwestern and eastern Wisconsin. Dry prairie associates include *Carex meadii*, *C. richardsonii*, side-oats grama (*Bouteloua curtipendula*), Hill's thistle (*Cirsium hillii*), bastard-toadflax (*Comandra umbellata*), white prairie-clover (*Dalea candida*), Robin's-plantain (*Erigeron pulchellus*), prairie alumroot (*Heuchera richardsonii*), yellow star-grass (*Hypoxis hirsuta*), violet woodsorrel (*Oxalis violacea*), hoary puccoon (*Lithospermum canescens*), little bluestem (*Schizachyrium scoparium*), bird's-foot violet (*Viola pedata*).

Similar species. Two other Wisconsin representatives of this section produce pistillate spikes at both the basal and upper nodes. *Carex tonsa* (shaved sedge [11]) grows almost exclusively in *dry sandy habitats* and has *perigynia 3–4.5 mm long*, larger than those of *C. umbellata*, with beaks generally ≥ 1 mm. *Carex deflexa* (northern oak sedge [9]) is sporadic in the northern highlands and central sands. It can be distinguished most easily from the preceding species by its *basal leaf sheaths, which do not become strongly fibrous as they age*, and by the lowest bract, which on elongated culms generally overtops the inflorescence.

Plant, inflorescence detail, and perigynium of *Carex umbellata*. Look closely for perigynia hidden among the leaves at the base of the plant.

15. *CAREX PENSYLVANICA*—PENNSYLVANIA SEDGE

section *Acrocystis*

After one of the localities from which the plant was described: "Cette plante croit dans la Pensylvanie, la nouvelle Yorck" (Mackenzie 1935, quoting Lamarck).

This widespread, early-flowering, strongly rhizomatous species with red, fibrous basal leaf sheaths, wispy foliage, and pubescent perigynia with globose bodies is one of eastern North America's most widely recognized sedges. Flowers April, fruits May to June, perigynia mostly falling in June.

Plants loosely cespitose or shoots arising singly; *rhizomes creeping, scaly, to > 0.25 m long; bases red-brown, clothed in fibrous sheaths of the previous year's leaves.* **Culms** 0.1–0.4 m tall. **Leaf blades** commonly equaling or exceeding the culms, mostly ≤ 3 mm wide. **Terminal spike** staminate, sessile or very short-stalked, 8–25 mm long. **Lateral spikes** pistillate, 1–3, sessile, congested near the base of the staminate spike but generally not overlapping, very short. **Perigynia** 2–3.2 mm long, 1–1.5 mm wide, prominently 2-ribbed, otherwise veinless, pubescent, *frequently blackened by fungal infection; body nearly spherical to obovoid,* base tapered, beak apex short-toothed.

Habitat and state range. Dry to dry-mesic woods and prairies, especially in sandy soils, ranging to mesic and bottomland forests throughout the state.

Similar species. Vegetative individuals of *Carex woodii* [30], *C. rosea* [117], and *C. radiata* [118] superficially resemble Pennsylvania sedge, but all are easily distinguished. The first two are *cespitose with rhizomes short or inconspicuous, bases not reddish or fibrous. Carex woodii,* on the other hand, has long-creeping rhizomes like *C. pensylvanica,* but its *burgundy-colored bases; wider, more erect leaves;* and *nonfibrous basal sheaths* distinguish it. *Carex communis* (colonial oak sedge [12]) is common in northern Wisconsin and occasional in southern mesic forests, the only species in the section with *widest leaf blades typically > 3 mm wide. Carex peckii* (Peck's sedge [16]), a northern species of open woodlands, has spikes crowded near the tip of the culm; staminate spikes 5–8.5 mm long, shorter than in *C. pensylvanica;* and perigynium bodies ellipsoid, yellowing at maturity. *Carex albicans* (oak sedge [18]) is largely confined to the central sands and has *ellipsoid perigynium bodies that are concealed by the pistillate scales. Carex inops* (sun sedge [14]) is found in scattered woodlands and fields of the Driftless Area and northeastern Wisconsin, often in sandy soils. Perigynia in this species are generally wider than those of Pennsylvania sedge (> 1.5 mm wide), as are the achenes (≥ 1.5 mm wide), but the two species are not easily distinguished.

Plant, inflorescence, and perigynium of *Carex pensylvanica*. The rhizomes, slender leaves, and fibrous bases are obvious throughout the year.

19. *CAREX LIMOSA*—MUCK SEDGE

section *Limosae*

Latin: mud.

Pendulous pistillate spikes and roots covered in a yellow, feltlike tomentum make this species and its relatives conspicuous in sphagnum bogs, conifer swamps, fens, and other peaty wetlands. Flowers May to June, fruits June, perigynia hanging on into fall.

Plants rhizomatous and stoloniferous, loosely cespitose or shoots arising singly; basal leaf sheaths purple to red; *roots clothed in a dense, yellow, feltlike tomentum* (may be pale in herbarium specimens). **Culms** often leaning or decumbent, 0.2–0.6 m tall, the previous year's culms often sending out shoots from the nodes. **Leaf blades** few, those of the sterile shoots arising mostly near the base of the culms, involute, tapering to a long, narrow apex, ≤ 2.5 mm wide. **Terminal spike** staminate, erect. **Lateral spikes** pistillate or occasionally androgynous (with a few male flowers at the apex), pendulous from slender stalks. **Pistillate scales** reddish brown to coppery, *as wide as the perigynia and concealing them,* apex acute to toothed. **Perigynia** whitish to pale green, 2.5–4 mm long, papillose distinctly veined on both faces, short-beaked.

Habitat and state range. Acidic bogs and peaty calcareous wetlands, primarily in the northern third of the state, with additional populations in calcareous wetlands of southeastern Wisconsin. Typical sphagnum bog associates include *Carex chordorrhiza, C. echinata, C. lasiocarpa, C. magellanica, C. oligosperma,* cottongrass (*Eriophorum virginicum*), sundews (*Drosera* spp.), bog lobelia (*Lobelia kalmii*). Typical calcareous sedge meadow associates include *C. buxbaumii, C. sterilis, C. tetanica,* and marsh muhly (*Muhlenbergia glomerata*).

Similar species. The *Limosae* are similar in perigynium and pistillate scale shape and coloration to **Carex buxbaumii** (**Buxbaum's sedge [26]**) and its relatives and to section *Paniceae;* however, none of these species possess the yellow roots of the *Limosae.* Wisconsin's other section *Limosae* species, *C. magellanica* (**boreal bog sedge [20]**), grows with *C. limosa* in sphagnum bogs of northern Wisconsin as well as occasional fens and white cedar swamps. *Carex magellanica* is recognizable by its *flat leaf blades ≤ 4 mm wide;* pistillate scales that are brown to blackish and *narrower than the perigynia, not concealing them;* and shorter staminate spikes. *Carex magellanica* is recognized in many treatments as *C. paupercula.* North American plants are *C. magellanica* ssp. *irrigua.*

Plant, inflorescence, and perigynium of *Carex limosa*. The margins of the perigynia are just visible in the inflorescence illustrated here, peeking out from behind the pistillate scales.

26. Carex buxbaumii—Buxbaum's sedge
section *Racemosae*
After the German botanist Johann Christian Buxbaum (1693–1730).

This species' dark, narrow scales and gynecandrous terminal spike, which suggests an ice cream cone topped with perigynia, are distinctive. Flowers May (to June in the north), fruits May to June, perigynia falling in July or August.

Plants loosely cespitose or shoots arising singly; rhizomes elongate; basal leaf sheaths reddish, fibrillose, bearing short, stiff blades. **Culms** 0.25–1 m tall, scabrous beneath the inflorescence. **Leaf blades** conspicuously bluish in the spring, 2–3.5 mm wide. **Terminal spike** *usually gynecandrous, the staminate scales forming a clavate base that flares outward at the base of the pistillate portion*, occasionally staminate or with staminate flowers at the apex and the base. **Lateral spikes** pistillate, 2–4, ascending, short-stalked. **Pistillate scales** *dark purple to brown with a pale midvein*, some longer than the perigynia, *tapering to a long-pointed apex*. **Perigynia** whitish to pale green, approximately 3 mm long, widest below the middle, distinctly papillose; beak short or lacking, shallowly toothed.

Habitat and state range. Common in calcareous prairies, sedge meadows, fens, and peaty soils overlying wet sand; primarily in the southeastern, central, and northeastern portions of the state, occasional near Lake Superior and in the Driftless Area. Typical associates include *Carex conoidea, C. interior, C. sartwellii, C. sterilis, C. stricta, C. tetanica*, white colic-root (*Aletris farinosa*), prairie alumroot (*Heuchera richardsonii*), yellow star-grass (*Hypoxis hirsuta*), bog lobelia (*Lobelia kalmii*), winged loosestrife (*Lythrum alatum*), shrubby cinquefoil (*Pentaphylloides floribunda*), prairie phlox (*Phlox pilosa* and *P. glaberrima*), mountain mint (*Pycnanthemum virginianum*), whip nut-rush (*Scleria triglomerata*), Riddell's goldenrod (*Solidago riddellii*), edible valerian (*Valeriana edulis*).

Similar species. The circumboreal **Carex media** (intermediate sedge [27]) is the only other Wisconsin member of the section, known from a talus slope in Grant County. Upper spikes of the *C. media* inflorescence are tightly clustered, unlike those of *C. buxbaumii*, its pistillate scales are shorter than the perigynia, and its perigynia are widest near the middle. Young, bluish shoots of *C. buxbaumii* suggest young shoots of **C. stricta** (tussock sedge [70]), which differs in having *basal leaf sheaths long-tapering, bladeless*, deeper red than those of *C. buxbaumii*. Tussock sedge forms *large hummocks,* unlike the 1- or few-stemmed tufts of Buxbaum's sedge.

Plant, inflorescence, and perigynium of *Carex buxbaumii*.

28. *CAREX EBURNEA*—BRISTLE-LEAF SEDGE
section *Albae*
Latin: resembling ivory.

The wiry leaf blades and mat-forming habit of this species are distinctive, and its unique tubular bracts, inconspicuous staminate spikes, and short perigynium beaks that are whitened at the apex make even inflorescence scraps easy to identify. Flowers April to May, fruits May to June, perigynia generally falling in August or September.

Plants colonial, forming dense sods; rhizomes elongate; basal leaf sheaths brown. **Culms** wiry, 0.1–0.3 m tall. **Leaf blades** *wiry, involute, often curling,* ≤ 1 mm wide. **Bracts** *reduced to bladeless sheaths, tawny, the margins translucent.* **Terminal spike** staminate, sessile, partly concealed within the sheath of the uppermost bract. **Lateral spikes** pistillate, 2–4 on conspicuous slender stalks. **Pistillate scales** shorter than to equaling the perigynia, cradling the perigynium bases. **Perigynia** 1.5–2 (mostly ≤ 2 mm) long, *green initially, usually becoming black with age,* distinctly few-veined; *beaks very short and often whitened at the tip.* Hermann (1940, in Deam's *Flora of Indiana*) notes that perigynia persist long past maturity, frequently hanging on even after the culms have fallen prostrate in winter.

Habitat. A calciphile, most Wisconsin populations growing in juniper glades of the Driftless Area, often atop rock outcrops or in rock crevices, or in white cedar swamps along the shore of Lake Michigan (primarily Door County). In northern Wisconsin the species also occurs in mesic to wet calcareous sites within boreal forests and along a few alkaline lakeshores of Oconto County with *Carex crawei,* showy lady's-slipper (*Cypripedium reginae*), and boreal bog orchid (*Platanthera dilatata*). Typical juniper glade associates include all three of Wisconsin's juniper species, dwarf cliff brake (*Pellaea glabella*), columbine (*Aquilegia canadensis*), rock cress (*Arabis lyrata*), side-oats grama (*Bouteloua curtipendula*), hairy grama (*Bouteloua hirsuta*), bluebell (*Campanula rotundifolia*), little bluestem (*Schizachyrium scoparium*). Typical white cedar swamp associates include paper birch, black ash, bulblet fern (*Cystopteris bulbifera*), big-leaved aster (*Aster macrophyllus*), yellow lady's-slipper (*Cypripedium calceolus*), wild lily-of-the-valley (*Maianthemum canadense*).

Similar species. The thin, wiry leaf blades and sparse inflorescence of this plant are utterly distinctive.

Plant, inflorescence, and perigynium of *Carex eburnea*, showing the bladeless bracts and inconspicuous staminate flowers.

30. *CAREX WOODII*—WOOD'S SEDGE, PRETTY SEDGE
section *Paniceae*
After William A. Wood, one of the plant's discoverers.

To fully appreciate the beauty of this early-blooming plant, let sunlight shine through the bladeless, red to purple basal leaf sheaths. The basal leaf sheaths and erect, narrow leaves are distinctive at a distance. Flowers May, fruits May to June, perigynia mostly falling in June and July.

Plants clonal, shoots arising singly and often covering large areas of the forest floor; rhizomes elongate, slender; *basal leaf sheaths red to purple, bladeless.* Culms 0.3–0.6 m tall. Leaves stiff, erect, mostly shorter than the culms, 2–4 mm wide. Staminate spike solitary, erect, generally elevated above the tops of the pistillate spikes. Pistillate spikes erect, loosely flowered, on slender stalks. Perigynia pale, 3–4 mm long, distinctly veined, tapered to the base; beak short, bent.

Habitat and state range. Rich, deciduous forests, most often floodplains of small rivers, scattered throughout the state but absent from most of south-central and far northwestern Wisconsin. Typical associates include sugar maple, basswood, hemlock, yellow birch, leatherwood (*Dirca palustris*), *Carex arctata, C. hirtifolia, C. gracillima, C. plantaginea, C. rosea, C. spren-gelii,* maidenhair fern (*Adiantum pedatum*), red baneberry (*Actaea rubra*), false rue anemone (*Enemion biternatum*), Virginia waterleaf (*Hydrophyllum virginianum*), wood nettle (*Laportea canadensis*), woodland phlox (*Phlox divaricata*), mayapple (*Podophyllum peltatum*), nodding trillium (*Trillium cernuum*).

Similar species. Although vegetative shoots of Wood's sedge are sometimes confused with *Carex pensylvanica* (Pennsylvania sedge [15]), the deeply colored, nonfibrous basal leaf sheaths and broader, stiffer leaves of Wood's sedge are distinctive. Wood's sedge is closely related to *C. meadii* (Mead's sedge [32]) and *C. tetanica* (common stiff sedge [33]), neither of which occurs in the forested habitats of *C. woodii.* Neither *C. meadii* nor *C. tetanica* has the purple basal leaf sheaths of Wood's sedge, though *C. tetanica* can be reddish-tinged at the base. Both have rhizomes that grow more deeply than those of Wood's sedge and basal leaf sheaths with blades (the plants are phyllopodic), whereas Wood's sedge is aphyllopodic.

Plant, pistillate spike, and perigynium of *Carex woodii*. Note especially the red, bladeless basal leaf sheaths, elongate rhizomes, and loosely flowered pistillate spike.

32. *CAREX MEADII*—MEAD'S SEDGE

section *Paniceae*

[= *Carex tetanica* var. *meadii*]; after its discoverer, Samuel Barnum Mead (1799–1880).

Firm, grayish foliage and pale yellow perigynia with raised veins and short, bent beaks distinguish most individuals of this relatively common dry prairie species. Flowers April to May, fruits May to early July, perigynia often falling by the end of July.

Plants clonal, sometimes loosely cespitose, shoots arising singly or few to a clump; rhizomes deep, elongate; basal leaf sheaths brown. **Culms** 0.1–0.7 m tall. **Leaf blades** *grayish green, firm,* margins often revolute, usually ≤ 4 (occasionally 7) mm wide. **Terminal spike** staminate, *long-stalked, elevated prominently above the lateral spikes.* **Lateral spikes** pistillate, erect to ascending, 0.5–3.5 cm long. **Pistillate scales** *brown to purplish* with translucent margins and a pale band flanking the midvein. **Perigynia** *pale yellowish green* to brown, especially later in the season, 3–4 mm long, strongly veined; beak short, bent.

Habitat and state range. Common in dry lime prairies, less common in sand prairies and oak barrens, southern two-fifths of the state. The plant is occasional in low prairies, where it is easily confused with *Carex tetanica.* Typical associates include *C. richardsonii, C. umbellata,* leadplant (*Amorpha canescens*), big bluestem (*Andropogon gerardii*), heath aster (*Aster ericoides*), side-oats grama (*Bouteloua curtipendula*), hairy grama grass (*Bouteloua hirsuta*), bastard-toadflax (*Comandra umbellata*), yellow star-grass (*Hypoxis hirsuta*), prairie dropseed (*Sporobolus heterolepis*), hoary puccoon (*Lithospermum canescens*), mountain mint (*Pycnanthemum virginianum*), little bluestem (*Schizachyrium scoparium*), Indian grass (*Sorghastrum nutans*).

Similar species. *Carex meadii* is easily confused with **C. *tetanica*** (**common stiff sedge** [33]), a species of wet prairies, fens, sedge meadows, and other open calcareous wetlands in the southern two-thirds of Wisconsin. *Carex tetanica* is absent from dry prairies, the more common habitat of *C. meadii* in Wisconsin. However, *C. meadii* occasionally grows in calcareous wet prairies or fens alongside *C. tetanica,* and in parts of its range it may actually be more common in these habitats. *Carex tetanica* typically has *softer green foliage* and *green perigynia.*

Plant, inflorescence, and perigynium of *Carex meadii*. The stiff appearance of the plant in this illustration is typical.

35. *CAREX CRYPTOLEPIS*—SMALL YELLOW SEDGE
section *Ceratocystis*
[= *Carex flava* var. *fertilis*]; Greek: hidden scale.

The squat, prickly yellow spikes of this species are unmistakable. No other sedge in our flora has perigynia so beautifully recurved as those of section *Ceratocystis*. Flowers May to June, fruits June, perigynia typically falling in September.

Plants cespitose; rhizomes short. **Culms** 5–70 cm tall. **Leaf blades** generally shorter than the culms, ≤ 4 mm wide. **Terminal spike** staminate or mostly staminate, often bearing some perigynia near the middle, typically short-stalked with its base nestled among the pistillate spikes. **Lateral spikes** pistillate or androgynous, *frequently tipped with a short plume of male flowers (looking like a cone of empty scales most of the season)*, short-stalked or the lower longer-stalked, prickly with perigynium beak tips, 1–2 times as long as thick. **Pistillate scales** the same color as the perigynia. **Perigynia** *arched away from the inflorescence axis, the lower reflexed, flattened,* yellow, approximately 4 mm long, strongly veined on both faces; beak smooth, roughly equaling the body in length, reflexed.

Habitat and state range. Wet sandy lakeshores or exposed sand, calcareous sedge meadows, and cedar swamps. *Carex cryptolepis* is found in four regions of Wisconsin: northwestern Wisconsin; Marquette, Waushara, Portage, and Waupaca counties; extreme northeastern Wisconsin; and Door County. The species also occurs in scattered wetlands of southeastern Wisconsin, including a population in the UW–Madison Arboretum. Although the species has been characterized as rare in calcareous soils of Canada and the northern United States, many Wisconsin collections are from calcareous wetlands. Associates include *C. aurea, C. bebbii, C. buxbaumii, C. crawei, C. granularis, C. viridula*, cottongrasses (*Eriophorum* spp.), northern bog bedstraw (*Galium labradoricum*), Kalm's Saint-John's-wort (*Hypericum kalmianum*), Canada rush (*Juncus canadensis*), hard-hack (*Spiraea tomentosa*), marsh fern (*Thelypteris palustris*), bog Saint-John's-wort (*Triadenum fraseri*).

Similar species. *Carex cryptolepis* closely resembles **C. *flava* (large yellow sedge [36])**, which differs in having *perigynium beaks scabrous at least near the tip* (15× magnification) and *fresh pistillate scales coppery brown* that are conspicuous until the perygynia reflex. Large yellow sedge is absent from northwestern Wisconsin and a stricter calciphile than *C. cryptolepis*; otherwise it is similar in ecology and distribution. **Carex *viridula* (little green sedge [34])** is similar to *C. flava* and *C. cryptolepis* in morphology and distribution but has *straight or slightly reflexed perigynia* with *beaks less than half as long as the body.*

Plant, inflorescence, and perigynium (back view and side view) of *Carex cryptolepis*. The combination of reflexed perigynia with bent beaks makes section *Ceratocystis* very easy to recognize.

38. *CAREX GRACILLIMA*—GRACEFUL SEDGE

section *Hymenochlaenae*

Latin: slender.

The combination of a gynecandrous terminal spike and beakless perigynia borne in slender, pendulous lateral spikes makes *Carex gracillima* easy to identify. This is one of southern Wisconsin's most common upland forest sedges. Flowers April to May, fruits May to June, perigynia falling in late June or early July.

Plants cespitose; *basal leaf sheaths purple.* **Culms** 0.2–1 m tall, usually leaning at maturity. **Leaf sheaths** glabrous, occasionally sparsely pubescent. **Leaf blades** dark green, 3–9 mm wide. **Terminal spike** *gynecandrous, predominantly staminate, tipped with perigynia,* 1–6 cm long. **Lateral spikes** pistillate, 2–5, mostly on the upper half of the culm, borne well above the foliage, at least the lower on *long, drooping stalks.* **Perigynia** 2–4 mm long, round-triangular in cross-section, veins approximately 10, base acute, *apex beakless.*

Habitat and state range. Common in mesic to bottomland forests throughout the state, ranging to oak woodlands and, especially northward, more open sites such as wet roadsides and ditches. Typical associates include red oak, sugar maple, *Carex arctata, C. deweyana, C. grisea, C. hirtifolia, C. intumescens, C. leptonervia, C. radiata, C. rosea, C. sprengelii,* jack-in-the-pulpit (*Arisaema triphyllum*), bearded shorthusk (*Brachyelytrum erectum*), blue cohosh (*Caulophyllum thalictroides*), woodland tick-trefoil (*Desmodium glutinosum*), lopseed (*Phryma leptostachya*), clearweed (*Pilea pumila*).

Similar species. *Carex gracillima* is readily distinguished from the other members of its section by the combination of gynecandrous terminal spikes, beakless perigynia, and leaf sheaths that are almost always glabrous. *Carex arctata* (**drooping woodland sedge [45]**) is common in forests of northern Wisconsin and the Baraboo Hills, where it replaces or accompanies *C. gracillima.* The species are superficially similar, but *perigynia of C. arctata taper to the beak and are constricted at the base to form a short stipe,* and its terminal spikes are staminate. *Carex debilis* (**northern weak sedge [46]**), relatively common in wet forests and woodlands of central and northern Wisconsin, is similar to *C. arctata,* but its leaves are narrower, and its *perigynia taper to a slender, acute base without a stipe.*

(a) Plant, (b) terminal and lateral spike, and (c) perigynium of *Carex gracillima;* (d) perigynium of *C. arctata.* The gynecandrous terminal spike of *C. gracillima* and several of its relatives is as obvious in the field as it is in this illustration. The difference in perigynium shape between *C. gracillima* and *C. arctata* and relatives is also obvious in the field, although the perigynium stipe of *C. arctata* is not always so obvious.

41. *Carex sprengelii*—Sprengel's sedge

section *Hymenochlaenae*

[= *Carex longirostris*]; after Kurt Sprengel (1766–1833), German physician and botanist.

The long-beaked perigynia of this species resemble the volumetric flasks of a chemist's laboratory. The long fibers at the base of the plant make this species recognizable even in midwinter, and the characteristic nodding of the leaves in midsummer may be visible from across a field. Flowers late April to May, fruits May to June, perigynia mostly falling by late June.

Plants densely cespitose; rhizomes stout, short, fibrillose; *base clothed in long brown fibers.* **Culms** 0.3–1 m tall, often leaning. **Leaves** 2–4 mm wide, stiffly erect at first, *tending to nod and then crease at about the midpoint in midsummer.* **Upper 1–4 spikes** staminate or androgynous, often with a few perigynia at the base or in the middle, ascending. **Lower spikes** pistillate, borne on long stalks, *drooping or the uppermost ascending.* **Pistillate scales** diverging at maturity, roughly equaling the perigynium beak tip, awned or tapering to a long, awl-like tip. **Perigynia** 4.5–6.5 mm long; *body globose, 2-ribbed, glossy; beak abrupt, cylindrical, slender, roughly equaling the body in length.*

Habitat and state range. Rich mesic to bottomland forests and forest edges, occasional oak woodlands; tolerant of disturbance, not flourishing in dense shade. Sprengel's sedge is most common in the southern third of Wisconsin, but populations occur in virtually every county. Common associates include *Carex hirtifolia, C. pensylvanica, C. radiata, C. woodii,* maidenhair fern (*Adiantum pedatum*), Dutchman's-breeches (*Dicentra cucullaria*), false rue anemone (*Enemion biternatum*), Virginia waterleaf (*Hydrophyllum virginianum*), bloodroot (*Sanguinaria canadensis*), twisted-stalks (*Streptopus* spp.), trillium (*Trillium grandiflorum* and *T. flexipes*).

Similar species. In fruit, this species is unmistakable. When identifying vegetative material based on the fibrous basal sheaths, don't confuse this species with **Carex pensylvanica** (**Pennsylvania sedge [15]**), which is readily distinguished by its reddish bases, *elongate rhizomes,* and much *finer foliage,* leaf blades commonly < 2 mm wide.

Plant, pistillate spike, and perigynium of *Carex sprengelii*.

47. CAREX ALBURSINA—WHITE-BEAR SEDGE

section *Laxiflorae*

Latin: white bear, after White Bear Lake, Minnesota, where the species was found in abundance by Edmond P. Sheldon, who elevated it to species level. It had previously been recognized as a variety of *C. laxiflora*.

White-bear sedge is southern Wisconsin's only common woodland and forest sedge with leaves more than 1 cm wide. Flowers April to May, fruits May, perigynia falling in June and July.

Plants cespitose, basal leaf sheaths pale to brownish, soft. **Culms** 0.1–0.6 m tall, weak, frequently leaning or decumbent, *angles narrowly winged.* **Leaf blades** soft, *1–4 cm wide* (widest leaves occasionally < 1 cm wide, in which case use inflorescence characters to distinguish the species from *Carex blanda*), the overwintering leaves conspicuous, at least in the early spring. **Upper bracts** *wider than the spikes, concealing them,* sheath margins smooth. **Terminal spike** staminate, *sessile or very short-stalked, the base approximately even with the base of the uppermost pistillate spike.* **Lateral spikes** pistillate, ascending, *loosely flowered,* the upper lateral spikes sometimes crowded. **Pistillate scales** blunt or obtuse. **Perigynia** 3–4.5 mm long, strongly many-veined, tapering or constricted to the base; beak short, bent away from the inflorescence axis.

Habitat and state range. Most common in mesic forests, southern two-thirds of the state; uncommon in sandier soils of central Wisconsin. White-bear sedge often persists in disturbed forests but does not tolerate disturbance to the extent that *Carex blanda* does. Typical associates include sugar maple, basswood, red oak, *C. hirtifolia, C. hitchcockiana, C. jamesii, C. oligocarpa, C. plantaginea, C. radiata, C. rosea, C. sparganioides,* maidenhair fern (*Adiantum pedatum*), wild leek (*Allium tricoccum*), white trout-lily (*Erythronium albidum*), goldenseal (*Hydrastis canadensis*), twinleaf (*Jeffersonia diphylla*), bloodroot (*Sanguinaria canadensis*).

Similar species. Narrow-leaved individuals may be confused with **Carex blanda (common wood sedge [50]),** but the wide bracts, blunt apex of the pistillate scales, and loosely flowered spikes of white-bear sedge distinguish it. Two other wide-leaved forest species occur in the state, both from section *Careyanae.* **Carex plantaginea (plantain-leaved sedge [59])** is relatively common in northeastern Wisconsin mesic forests, sporadic in northern and west-central Wisconsin, and most easily distinguished by its *purple bases; dark green leaf blades that are strongly W-shaped in cross-section and puckered on the margins; and purple-sheathed, short-bladed bracts.* **Carex platyphylla (broad-leaf sedge [61])** inhabits sugar maple–beech forests of Door County, where its *slender culms* (frequently less than 0.5 mm thick) and *grayish to blue-green, strongly glaucous leaves* distinguish it.

Carex albursina, whole plant. Note the pale base, broad leaves, and broad bracts, which distinguish this species from everything else in the flora.

50. *CAREX BLANDA*—COMMON WOOD SEDGE
section *Laxiflorae*
Latin: soft, smooth, caressing.

The pale, soft bases of this common plant feel almost rubbery in spring. By early summer the leaves and culms are usually sprawling on the forest floor. Flowers May, fruits May to June.

Plants cespitose, bases pale to brownish. **Culms** 0.1–0.6 m tall, *narrowly winged and scabrous on the angles.* **Leaf blades** green to glaucous, ≤ 1 cm wide. **Bract sheaths** *scabrous along the margins.* **Terminal spike** staminate, 1–2 cm long, *sessile or short-stalked (peduncle to 1.5 cm long), often closely adjacent to the uppermost pistillate spike.* **Lateral spikes** pistillate, emerging mostly from the upper half of the culm, the upper lateral spikes crowded together, occasionally one long-stalked pistillate spike emerging from a lower sheath. **Pistillate scales** whitish-hyaline, the green midrib extended to the apex to form a tooth or short, scabrous awn. **Perigynia** 2.5–4 mm long, strongly 25–32-veined; beak short, bent away from the inflorescence axis, in profile reminiscent of the head of a chick.

Habitat and state range. Common in mesic to wet deciduous forests in the southern two-thirds of the state, especially in wet microsites and silt loam soils; ranging to mesic savannas, brushy thickets, prairies, garden beds, sidewalk cracks, and roadsides; highly tolerant of disturbance. *Carex blanda* is somewhat weedy but native and not invasive. Typical associates include sugar maple, red oak, basswood, *C. communis, C. gracillima, C. rosea, C. sparganioides,* hog-peanut (*Amphicarpaea bracteata*), woodland tick-trefoil (*Desmodium glutinosum*), wild geranium (*Geranium maculatum*), bristly buttercup (*Ranunculus hispida*), false Solomon's-seal (*Smilacina racemosa*).

Similar species. *Carex leptonervia* (**few-nerved wood sedge** [49]) is nearly as common in northern forests as *C. blanda* is in the south. It differs from all other members of the section in having *perigynia with 2–3 very distinct veins,* the remaining veins subtle or inconspicuous. See the key for other species in the section (pp. 37–38) as well as the discussion under *C. albursina* (**white bear sedge** [47]). The unrelated *C. grisea* (**gray sedge** [56]) is vegetatively similar and grows in similar habitats but is more robust with *green bases; firm, narrower, somewhat glossy leaf blades;* and slender, firm culms.

Plant, inflorescence, and perigynium of *Carex blanda*. The shoots are weak and often more reclining in the field.

54. *CAREX GRANULARIS*—LIMESTONE MEADOW SEDGE

section *Granulares*

Latin: grainlike.

Basal rosettes of blue-gray, glaucous leaves splayed on the ground in calcareous wetlands often make this plant recognizable even before it flowers. Flowers May, fruits May to June, perigynia sometimes persisting through August.

Plants cespitose; rhizomes short. **Culms** ≤ 0.6 m tall. **Leaf blades** *glaucous, especially near the base of the blade,* W-shaped in cross-section, the widest typically > 5 mm wide. **Terminal spike** staminate, sessile or short-stalked, the base overlapped by the uppermost pistillate spike(s). **Lateral spikes** pistillate, *narrowly oblong.* **Pistillate scales** *flecked with red dots or streaks,* especially late in the season (this can be seen with a 10× hand lens under direct light). **Perigynia** > 25 per spike, *flecked with red,* 2–3 mm long, distinctly veined, short-beaked.

Habitat. Calcareous wet forests to moderately shady wet areas; occasional on gravel roadsides, calcareous wet prairies and fens or boglike areas, very occasional in mesic forests. The plant is most common in the eastern third of Wisconsin, though a few populations are found along the far western and northern borders of the state. Typical associates include *Carex aurea, C. buxbaumii, C. eburnea, C. flava, C. interior, C. pellita, C. sterilis, C. tetanica, C. viridula,* redtop (*Agrostis gigantea*), spring-cress (*Cardamine bulbosa*), showy lady's-slipper (*Cypripedium reginae*), Kalm's Saint-John's-wort (*Hypericum kalmianum*), pale spike lobelia (*Lobelia spicata*), grass-of-Parnassus (*Parnassia glauca*), prairie phloxes (*Phlox pilosa* and *P. glaberrima*), black bulrush (*Scirpus atrovirens*).

Similar species. The closest relative in Wisconsin is *Carex crawei* (**early fen sedge [53]**), a species of wet, open, calcareous habitats in eastern Wisconsin distinguished by its *elongate rhizomes, narrower leaves,* and *long-stalked staminate spike.* In fruit, *C. granularis* superficially resembles *C. blanda* (**common wood sedge [50]**), which differs in having *weak culms, strongly bent perigynium beaks,* and *shorter pistillate spikes.* Pistillate spikes of our *Granulares* species superficially resemble those of *C. pallescens* (**pale sedge [23]**), a species of wet ditches and occasional wet to mesic forests in a few counties of far north-central and southeastern Wisconsin that has *beakless perigynia* and *pubescent culms, leaf sheaths, and leaf blade undersides.*

Plant, inflorescence, pistillate scale, and perigynium of *Carex granularis*. In the field with a 10× hand lens, pigmentation in the pistillate scales will be about as obvious as it is in this illustration. With better magnification and lighting the pigmentation forms discrete, obvious flecks.

55. *Carex conoidea*—Prairie gray sedge

section *Griseae*

Greek: cone-shaped.

This species is not frequently collected in Wisconsin, but the plant appears to be relatively common in wet prairies. Flowers mid- to late May, fruits late May to June, perigynia mostly falling in July and August.

Plants cespitose; basal leaf sheaths yellow to brown. **Culms** 0.15–0.5 m tall, generally < 1 mm thick, angles scabrous. **Leaf blades** not as long as the culm, the widest usually 3–4 mm wide. **Terminal spike** staminate, *elevated well above the pistillate spikes* on a scabrous stalk, usually equaling or exceeding the uppermost bract. **Lateral spikes** pistillate, usually 1–2 near the tip of the culm on scabrous stalks, 1 or more sessile or slenderly stalked lower on the plant. **Perigynia** slightly inflated, yellow, roughly conical in shape (shaped a bit like an old-fashioned Coke bottle), 2.5–4 mm long, tapering to a blunt base and a short or inconspicuous beak, *lustrous, veins 16–20, impressed (sometimes raised in immature material).*

Habitat. Most characteristic of wet calcareous prairies, frequently persisting in calcareous old fields, ranging occasionally to sedge meadows and sphagnous jack pine woodlands; primarily in southern and central Wisconsin, a few collections from wetlands along the west edge of Green Bay, one collection from Burnett County. Typical wet prairie associates include *Carex buxbaumii, C. tenera, C. tetanica*, marsh fern (*Thelypteris palustris*), New England aster (*Aster novae-angliae*), shooting-star (*Dodecatheon meadia*), blunt-leaf bedstraw (*Galium obtusum*), prairie alumroot (*Heuchera richardsonii*), Kalm's Saint-John's-wort (*Hypericum kalmianum*), yellow star-grass (*Hypoxis hirsuta*), prairie blazing-star (*Liatris pycnostachya*), winged loosestrife (*Lythrum alatum*), stiff cowbane (*Oxypolis rigidior*), northern ragwort (*Packera paupercula*), shrubby cinquefoil (*Pentaphylloides floribunda*), prairie phloxes (*Phlox pilosa* and *P. glaberrima*), white goldenrod (*Solidago ptarmicoides*), Riddell's goldenrod (*S. riddellii*), stiff goldenrod (*S. rigida*).

Similar species. Learn to recognize the unique perigynium surface in prairie gray sedge—lustrous with impressed veins—and you will be unlikely to confuse it with anything else. *Carex tetanica* (common stiff sedge [33]) is a wet prairie and fen species similar in stature and appearance to *C. conoidea*, but with *rhizomes elongate (plants not cespitose), perigynium veins raised,* and *perigynium beaks short, bent.*

Plant, pistillate spike, and perigynium of *Carex conoidea.*

56. *Carex grisea*—Gray sedge

section *Griseae*

[= *Carex amphibola* var. *turgida*]; Latin: gray.

Gray sedge often forms robust, dark green clumps along moist, shaded trail edges. Its perigynia are among the largest of Wisconsin's upland forest species. Flowers May, fruits May to June, perigynia mostly falling in July.

Plants densely cespitose. **Culms** 0.2–0.7 m tall, smooth. **Leaf blades** glossy on the upper surface, 4–8 mm wide. **Bracts** wide, foliose, the lowest equaling or exceeding the tip of the inflorescence. **Terminal spike** staminate, *sessile, its base nestled beside the base of the uppermost pistillate spike,* generally exceeding the latter. **Lateral spikes** pistillate, loosely flowered, often with ≤ 10 perigynia (occasionally as many as 20); the upper two lateral spikes may be separate or paired, clustered at the tip of the plant, while the rest are more distant and sometimes drooping on slender stalks. **Pistillate scales** whitish-hyaline, midveins green, extending into short but prominent awns that may equal or exceed the perigynium apex, especially the lower perigynia in each spike. **Perigynia** slightly inflated, *nearly terete, roughly football-shaped at first,* becoming rounded and broader at the tip, 4.5–5.5 mm long; *veins > 50, fine, impressed (often raised in immature material).*

Habitat and state range. Most common in mesic to wet deciduous forests, ranging to wet roadsides and ditches; primarily southern Wisconsin, mostly absent from the central sands, ranging north of the Tension Zone to Sheboygan, Manitowoc, and Brown counties; also known from the banks of the Menominee River in the Upper Peninsula of Michigan. Gray sedge is tolerant of disturbance and often thrives along trails. Typical associates include sugar maple, red oak, slippery elm, *Carex blanda, C. gracillima, C. radiata, C. woodii.*

Similar species. *Carex grisea* can be confused before its perigynia are ripe with *C. blanda* (**common wood sedge** [50]), a less robust plant that grows in similar wet-mesic forests but has softer foliage, *weak, scabrous wing-margined culms,* and pistillate scales that are somewhat shorter than the perigynia. See also the key to section *Griseae* species (p. 38).

Plant, inflorescence, and perigynium of *Carex grisea*.

62. *CAREX LAXICULMIS*—WEAK-STEMMED WOOD SEDGE

section *Careyanae*

Latin: loose-stemmed.

The lax habit and exceedingly narrow, almost threadlike stalks of the lateral spikes make this uncommon mesic forest species distinctive. Flowers May, fruits June.

Plants cespitose; basal leaf sheaths brown. **Culms** 0.1–0.5 m tall, lax, angles not winged or scabrous. **Leaf blades** glaucous (var. *laxiculmis*) to bright green (var. *copulata*), the widest mostly ≤ 9 mm (occasionally > 1 cm). **Terminal spike** staminate, stalk very short to roughly as long as the uppermost pistillate spike. **Lateral spikes** pistillate, loosely flowered; lowest spike emerging from the lowest sheath on the culm, *pendulous on a narrow stalk; lowest scale of each pistillate spike sterile (empty) or subtending a staminate flower.* **Perigynia** *acutely triangular in cross-section,* 2.5–4 mm long, finely veined; beak short, bent.

Habitat and state range. Sugar maple forests; scattered localities primarily in the southern half of the state, with two collections of *Carex laxiculmis* var. *laxiculmis* in Menominee and Marinette counties.

Similar species. The two varieties overlap morphologically: *Carex laxiculmis* var. *laxiculmis* has leaf blades typically glaucous, the widest ≥ 6.5 mm wide; *C. laxiculmis* var. *copulata*, the more widespread of the two, has leaf blades usually green, the widest ≤ 8.5 mm wide. *Carex laxiculmis* is closely related to *C. digitalis* (**slender woodland sedge [63]**), known from four collections in mesic forests of southeastern Wisconsin. *Carex digitalis* is easily distinguished by its *narrower leaf blades* (generally ≤ 5 mm wide) and the *pistillate flower in the lowest scale of each lateral spike.* See also *C. blanda* (**common wood sedge [50]**) and relatives.

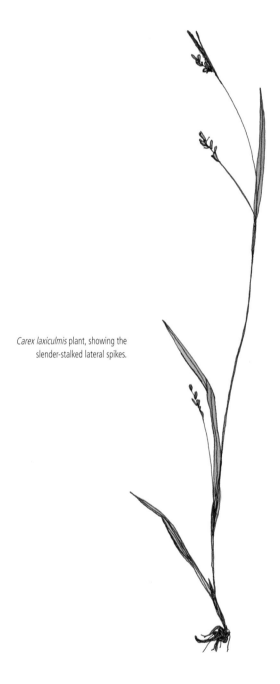

Carex laxiculmis plant, showing the
slender-stalked lateral spikes.

64. *CAREX AUREA*—GOLDEN SEDGE
section *Bicolores*
Latin: golden.

The fleshy perigynia of this species turn orange when they are about to fall from the plant, perhaps an adaptation to bird dispersal, though there appear not to be any published accounts of animal dispersal. The perigynia spread as they ripen, making their attachment to the plant at maturity appear precarious. Flowers May, fruits June.

Plants colonial, culms arising singly or in small clumps; rhizomes slender. **Culms** 5–35 cm tall. **Leaf blades** usually shorter than to equaling the culm, ≤ 3 mm wide. **Lowest bract** foliose, overtopping the inflorescence. **Lateral spikes** pistillate, lax, especially the lowermost, the lowermost perigynia of the lowermost spikes not overlapping. **Terminal spike** usually staminate, sometimes gynecandrous. **Pistillate scales** brown, midvein pale, margins translucent, apex abruptly narrowed to a short tip. **Perigynia** *pale at first, becoming orange and fleshy at maturity, roughly spherical or at least rounded at the apex,* 2–3 mm long, strongly veined, beakless. **Stigmas** 2.

Habitat and state range. Most common in calcareous wetlands, often in disturbed areas and open communities without strong competition; counties adjacent to Lake Michigan and Lake Superior, scattered collections inland to Juneau, Waupaca, Shawano, and Florence counties. One population appeared in 1997 in marl that had been disturbed by boardwalk construction in Gardner Marsh at the UW–Madison Arboretum (Dane County). Associates in wet meadows, fens, and ditches include *Carex granularis, C. leptalea, C. sterilis, C. viridula,* grass pink (*Calopogon tuberosus*), grass-leaved goldenrod (*Euthamia graminifolia*), Dudley's rush (*Juncus dudleyi*), pale-spike lobelia (*Lobelia spicta*). Associates in white cedar swamps include *C. crawei,* green twayblade (*Liparis loeselii*), boreal bog orchid (*Platanthera dilatata*).

Similar species. The distinctive shape and color of the perigynia make it difficult to confuse golden sedge with any other species. The only close relative in the state is **C. garberi** (**elk sedge** [65]), a calciphile from Door County, distinguished from golden sedge by its *whitish, ascending perigynia, densely flowered spikes,* and *typically gynecandrous terminal spike.*

Plant and inflorescence of *Carex aurea*. Perigynium color in the herbarium specimen illustrated here does not accurately represent the yellow-orange color of ripe perigynia in the field. The colonial tufts joined by slender rhizomes are characteristic.

66. *CAREX CRINITA*—FRINGED SEDGE

section *Phacocystis*

Latin: hairy.

The long-drooping pistillate spikes of this species and its relatives, bristly with scabrous-awned scales, are unmistakable. Flowers May to June, fruits June to July, ripening July and August; perigynia often fall off abruptly upon ripening but may persist through August.

Plants *densely cespitose, forming large clumps;* basal leaf sheaths red, becoming ladder-fibrillose. **Culms** 0.2–1.3 m tall, roughened. **Leaf sheaths** frequently tinged *brown to reddish coppery on the inner band,* smooth. **Leaf blades** frequently with 3 very strong veins and many finer veins. **Upper 1–3 spikes** staminate or occasionally gynecandrous, tipped with a few perigynia, rarely with a few perigynia at the base. **Lateral spikes** mostly pistillate, occasionally androgynous, tipped with a few staminate flowers, *elongate, drooping or pendulous.* **Pistillate scale bodies** *notched or truncate at the apex; awns scabrous, exceeding the perigynium apex.* **Perigynia** > 100 per spike, flattened at first and becoming inflated, nearly terete at maturity, 2–3.5 mm long, widest at or above the middle, strongly 2-ribbed but otherwise most often unveined or with a single vein extending the length of the achene in addition to the ribs; apex blunt or obtuse. **Achene** indented on the side or otherwise contorted. **Stigmas** 2.

Habitat. Typically in floodplain forests, black ash swamps, or wet depressions within otherwise mesic forests, ranging to marshes, sedge meadows, and shrub carrs along streams and rivers throughout the northern two-thirds of the state and the lower Wisconsin River valley. Typical associates include black ash, silver maple, *Carex hystericina, C. lupulina, C. projecta, C. retrorsa, C. stricta, C. tuckermanii, C. typhina,* bluejoint grass (*Calamagrostis canadensis*), reed manna grass (*Glyceria grandis*), fringed loosestrife (*Lysimachia ciliata*), moneywort (*L. nummularia*), marsh bluegrass (*Poa palustris*), American bur reed (*Sparganium americanum*).

Similar species. *Carex gynandra* (nodding sedge [67]) is considered by some taxonomists to be a variety of *C. crinita.* However, *C. gynandra* has *lower leaf sheaths scabrous, pistillate scale body apex acute or truncate, not notched,* and *perigynium apex acute rather than obtuse or rounded.* The two species share almost identical ecological and geographic ranges; morphological intermediates between the two occur occasionally.

Plant and perigynium of *Carex crinita;* pistillate spike detail of *C. gynandra.*

70. *CAREX STRICTA*—TUSSOCK SEDGE

section *Phacocystis*

Latin: drawn together, tight, rigid.

This species is most easily recognized in seasonally flooded habitats, where it forms prominent tussocks that are distinctive at all seasons. Flowers May to June, fruits June to July, perigynia mostly falling by mid-August.

Plants densely cespitose; *rhizomes elongate, spreading both laterally and vertically, the vertical rhizomes contributing to the formation of tall, peaty tussocks in wet sites.* **Lower leaf sheaths** red, *conspicuously ladder-fibrillose, scabrous.* **Leaf blades** 4–6 mm wide, ligule longer than wide. **Culms** 0.3–1.7 m tall. **Lowest bract** shorter than to roughly equaling the tip of the inflorescence. **Upper 2–3 spikes** staminate, erect. **Lower 3–4 spikes** *androgynous with a short staminate section at the apex*, sometimes pistillate, erect, densely packed with perigynia. **Pistillate scales** shorter than the perigynia, apex obtuse or acute. **Perigynia** *ovate, pale, not inflated,* 2–3.5 mm long, *tapering to the apex*, short-beaked to beakless with a short stipe.

Habitat and state range. See facing page.

Similar species. Three close relatives of tussock sedge are common in Wisconsin. *Carex haydenii* (**long-scaled tussock sedge** [71]) is a wet prairie species that has basal leaf sheaths reddish, ladder-fibrillose (as in *C. stricta*) or not; *pistillate scales narrow, acute at the apex, and longer than the perigynia;* and perigynia slightly inflated, rounded at the apex. *Carex aquatilis* (**water sedge** [68]) is common in wetlands throughout the state, generally where there is standing water for most of the year. Its *rhizomes often form an important component of floating peat mats.* It does not form tussocks. Unlike *C. haydenii*, the lowest bracts of *C. aquatilis* are *conspicuously longer than the inflorescence; basal leaf sheaths are pinkish* and *not fibrillose; leaves are often blue-green, glaucous;* and *perigynia are widest above the middle.* *Carex emoryi* (**Emory's sedge** [73]), a floodplain species that is most common in the southern three-quarters of the state, has basal leaf sheaths not fibrillose; the *inner band of the leaf sheaths convex at the summit; ligules wider than long, nearly truncate rather than peaked at the apex; leaves approximately twice as wide as those of C. stricta;* and perigynia veined.

Two less common species show up in the state: *Carex lenticularis* [69] of the Apostle Islands and Ashland County, which has *bracts conspicuously longer than the inflorescences* and *perigynia clearly veined;* and *C. nigra* [72] of wetlands in the Superior-Duluth area, which has *basal leaf sheaths not fibrillose; pistillate scales black with a light midrib;* and *perigynia dark-spotted on the upper half.* The Great Plains and western North American species *C. nebrascensis* (not treated elsewhere in this book) was collected in Milwaukee County in 1938 and is otherwise unknown from the state. It is the only *C. stricta* relative in the Midwest flora with *perigynium beaks prominently bidentate.*

Habitats of Wisconsin's common section *Phacocystis* species — *Carex stricta* group only

Carex aquatilis. Pond edges, riverbanks, marshes, lakeshores, bogs, ditches, typically in deeper water than its relatives; scattered throughout the state, less common in the Driftless Area.

Carex emoryi. A range of alluvial wetlands, including wet prairies, standing water or muddy sloughs, lowland forests (frequently within openings), and streambanks or sandbars; more common in the southern two-thirds of the state, scattered throughout much of the north. Common associates include silver maple, river birch, buttonbush, bluejoint grass (*Calamagrostis canadensis*), sensitive fern (*Onoclea sensibilis*), stinging nettle (*Urtica dioica*).

Carex haydenii. Common in marshes, sedge meadows, wet prairies, and wet waste areas, frequently in sandy soils; occasional in rich forests, wet areas within otherwise dry woodlands, or alder thickets in the north; ranges mostly south of a line from Milwaukee to Oshkosh to Stevens Point and the Black River. Associates include *C. bicknellii, C. conoidea, C. tenera* var. *tenera,* New England aster (*Aster novae-angliae*), bluejoint grass (*Calamagrostis canadensis*), bastard-toadflax (*Comandra umbellata*), shooting-star (*Dodecatheon meadia*), Quaker-ladies (*Houstonia caerulea*), yellow star-grass (*Hypoxis hirsuta*), prairie phlox (*Phlox pilosa*), Jacob's ladder (*Polemonium reptans*), blue-eyed-grass (*Sisyrinchium campestre*), grass-leaved goldenrod (*Euthamia graminifolia*), golden alexanders (*Zizia* spp.).

Carex stricta. Sedge meadows, fens, very wet prairies, lakeshores, ditches, and a variety of open wetlands throughout the state. Typical associates include *C. buxbaumii, C. comosa, C. diandra, C. hystericina, C. prairea, C. sartwellii, C. trichocarpa,* and a wide range of other wetland species.

(a) Plant, (b) inflorescence, and (c) perigynium of *Carex stricta*. (d) Inflorescence and (e) lateral spike of *C. aquatilis*.
(f) Lateral spike and (g) perigynium of *C. haydenii*. Note the long, narrow pistillate scales, which exceed the inflated perigynia.

d   *e*   *f*   *g*

75. *CAREX COMOSA*—BRISTLY SEDGE

section *Vesicariae*

Latin: covered with hair.

The nodding, bristly, bottlebrush-like pistillate spikes of *Carex comosa* are common in wetlands throughout the Midwest. Flowers May to June, fruits mid- to late June, perigynia mostly falling in mid-July and August.

Plants cespitose, often forming large clumps; rhizomes short; basal leaf sheaths brown. **Culms** 0.5–1.5 m tall, triangular in cross-section, generally scabrous beneath the inflorescence. **Leaf blades** W-shaped in cross-section, ≤ 17 mm wide; *sheaths and lower sections of broad leaf blades conspicuously septate-nodulose;* ligule longer than wide. **Terminal spike** staminate or occasionally androgynous, a few perigynia near the base. **Lateral spikes** pistillate or occasionally androgynous, plumply cylindrical, 1.2–2 cm thick, longer spikes frequently arching, the upper ascending, the lower generally on long, pendulous stalks. **Pistillate scales** narrow, midvein extending to form an elongate, scabrous awn, only the awn visible in the spike at maturity. **Perigynia** *reflexed, scarcely inflated,* packed densely in the spike, *closely enveloping the achene,* 4.5–8.5 mm long, conspicuously veined; *beak teeth spreading or curving outward, 1.2–2 mm long.*

Habitat and state range. Marshes, bogs, shallow water near lakeshores, and standing water in a wide variety of open wetlands; throughout the state but not widespread in the Driftless Area. Associates include *Carex aquatilis, C. atherodes, C. hystericina, C. lacustris, C. lasiocarpa,* bluejoint grass (*Calamagrostis canadensis*), water-arum (*Calla palustris*), marsh bellflower (*Campanula aparinoides*), marsh cinquefoil (*Comarum palustre*), water-willow (*Decodon verticillatus*), manna grasses (*Glyceria* spp.), northern water-horehound (*Lycopus uniflorus*), marsh bluegrass (*Poa palustris*), tear-thumb (*Polygonum sagittatum*), common arrowhead (*Sagittaria latifolia*), wool-grass (*Scirpus cyperinus*), bur reeds (*Sparganium* spp.), marsh fern (*Thelypteris palustris*), cattails (*Typha* spp).

Similar species. Bristly sedge may be confused with *Carex hystericina* (porcupine sedge [78]), which differs in its *reddish basal leaf sheaths* and *perigynia spreading to ascending, not reflexed, with straight, shorter beak teeth.* Apparent hybrids between these two species are rare. More similar to bristly sedge is *C. pseudocyperus* (cypresslike sedge [76]), a species of northern wetlands that differs in having *spikes more slender, perigynia shorter* (usually ≤ 6 mm long), and *beak teeth straight or only slightly curved, ≤ 1.2 mm long.*

Vegetative shoot, inflorescence, and perigynium of *Carex comosa*. The thick, often arching lateral spikes are typical, as are the curved perigynium beak teeth.

78. CAREX HYSTERICINA—PORCUPINE SEDGE

section *Vesicariae*

Greek: porcupine. Often this species is spelled "*Carex hystricina*" to reflect the etymology of the name, but the name was validly published as "*Carex hystericina.*"

This morphologically variable bottlebrush sedge is common in a wide range of calcareous wetlands. Flowers May to June, fruits mid- to late June, perigynia mostly falling mid-July through August.

Plants cespitose; rhizomes short; *basal leaf sheaths generally reddish to purple, slightly fibrillose.* **Culms** 0.1–1 m tall, sharply triangular in cross-section, angles scabrous beneath the inflorescence. **Leaf blades** W-shaped in cross-section, ≤ 10 mm wide, ligule as wide as long. **Terminal spike** staminate. **Lateral spikes** pistillate, occasionally androgynous, *generally ascending or arching*, the lower spikes frequently pendulous on elongate stalks, oblong to narrowly oblong, occasionally ovate-spherical, ≤ 4 cm long, approximately 1 cm thick. **Pistillate scales** lanceolate, 0.5–1 mm wide, whitish-hyaline, midrib green, extending to form a *scabrous awn*, only the awn visible in the spike at maturity. **Perigynia** *ascending, slightly inflated*, densely packed in the inflorescence, 4.5–7 mm long, ≤ 2 mm wide, conspicuously 13–20-veined; beak 2–3 mm long, slender; *beak teeth straight, < 1 mm long.*

Habitat and state range. Common throughout the state in fens, sedge meadows, white cedar swamps, and a variety of calcareous wetlands; occasional in very wet prairies. Typical associates include *Carex bebbii, C. buxbaumii, C. interior, C. leptalea, C. pellita, C. prairea, C. stipata, C. stricta, C. tetanica, C. vulpinoidea,* marsh marigold (*Caltha palustris*), marsh bellflower (*Campanula aparinoides*), turtlehead (*Chelone glabra*), bulblet water-hemlock (*Cicuta bulbifera*), water-horehounds (*Lycopus* spp.), narrow-leaved loosestrife (*Lysimachia quadriflora*), marsh fern (*Thelypteris palustris*).

Similar species. Porcupine sedge hybridizes occasionally with other members of section *Vesicariae,* but hybrids are sterile (achenes fail to develop properly). Two other relatively common bottlebrush species in the state have the combination of inflated perigynia with short, straight teeth. **Carex lurida** (shallow sedge [79]) grows in floodplains of the upper Wisconsin, Black, and LaCrosse rivers. It suggests a robust *C. hystericina* with *pistillate spikes > 1.5 cm thick* and *perigynia strongly inflated, > 2.5 mm wide, lustrous.* **Carex retrorsa** (retrorse sedge [82]) grows in floodplains, wet forests, and other wetlands throughout the state; it differs in having *pistillate scales neither awned nor scabrous* and *perigynia generally reflexed.*

Plant, inflorescence, and perigynium of *Carex hystericina*. Compare the perigynium with that of *C. comosa* [75]; the beak teeth alone distinguish the two species.

80. *CAREX OLIGOSPERMA*—FEW-SEEDED HOP SEDGE, WIREGRASS
section *Vesicariae*
Greek: few-seeded.

The few-flowered pistillate spike nestled into the axil formed by the bract and the long stalk of the staminate spike makes this species very easy to recognize. The species is common in bogs of the upper Midwest and most of northeastern North America. Flowers May to June, fruits June to July, perigynia persisting into August or September.

Plants loosely cespitose or shoots arising singly; rhizomes long; basal leaf sheaths red. Culms 0.3–1 m tall, slender. Leaf blades *wiry, involute, firm.* Bract stiff, straight, wiry, approximately equaling the tip of the staminate spike. Terminal spike staminate, elevated well above the pistillate spikes. Lateral spikes pistillate, 1–3. Perigynia 3–15 per spike, inflated, 4–6.5 mm long, leathery, shiny, distinctly veined; beak abrupt, short, with two short teeth.

Habitat and state range. Primarily in bogs; also in northern sedge meadows and other peaty wetlands of northern Wisconsin. *Carex oligosperma* and *C. lasiocarpa* are important components of floating bog mats, which are woven together in large part with their rhizomes. Associates include larch, black spruce, sweet gale (*Myrica gale*) and common bog ericads (e.g., leatherleaf, wild cranberry, Labrador-tea), *C. canescens*, *C. chordorrhiza*, *C. diandra*, *C. disperma*, *C. echinata*, *C. lasiocarpa*, *C. leptalea*, *C. limosa*, *C. magellanica*, *C. rostrata*, *C. stricta*, *C. tenuiflora*, *C. trisperma*, *C. utriculata*, bluejoint grass (*Calamagrostis canadensis*), wild calla (*Calla palustris*), grass-pink (*Calopogon pulchellus*), sundews (*Drosera* spp.), three-way sedge (*Dulichium arundinaceum*), cottongrasses (*Eriophorum* spp.), pitcher-plant (*Sarracenia purpurea*), arrow-grass (*Scheuchzeria palustris*), false mayflower (*Smilacina trifolia*), and other common bog species.

Similar species. Fertile individuals are unmistakable. Vegetative shoots could be confused with **Carex lasiocarpa** (wiregrass [89]), the other "wiregrass," which has similar wiry leaf blades and grows in all habitats where *C. oligosperma* is commonly found (though *C. oligosperma* is confined to acidic soils). The basal leaf sheaths of both species are reddish to purple, but the *basal sheaths of* C. lasiocarpa *are more strongly fibrillose*, and the *inside of its leaf sheaths have distinctive raised septa between the veins* that are lacking in *C. oligosperma*; peel back the leaf sheath and use a hand lens to see this.

Plant, pistillate spike, and perigynium of *Carex oligosperma*. The plump, short pistillate spikes of this species are utterly distinctive in the section.

81. *CAREX TUCKERMANII*—TUCKERMAN'S SEDGE

section *Vesicariae*

After Edward Tuckerman (1817–1886), American botanist and lichenologist.

Large clumps of Tuckerman's sedge grace Wisconsin's floodplain forest understories and the edges of wet depressions in northern forests. The inflated perigynia ascending and overlapping in vertical rows on arching, drooping, or pendulous cylindrical pistillate spikes are distinctive. Flowers May, fruits June, perigynia generally hanging on into September or October.

Plants cespitose; rhizomes short; basal leaf sheaths red to purple, ladder-fibrillose. **Culms** 0.1–1.3 m tall, frequently leaning, sharply triangular in cross-section, the angles scabrous beneath the inflorescence. **Leaf blades** dark green, ≤ 5 mm wide. **Upper 1–3 (usually 2) spikes** staminate or occasionally androgynous, a few perigynia at base. **Spikes** pistillate, oblong to thick-cylindrical, the upper short-stalked or sessile, the *lower generally drooping or pendulous on elongate stalks.* **Pistillate scales** awnless, margins entire. **Perigynia** *inflated,* 7.5–12.5 mm long, thin-walled, *shiny,* strongly veined; beak short-toothed, roughly one-third as long as the body. **Achenes** asymmetric, *indented or invaginated on one side.*

Habitat and state range. Floodplain forests, unshaded sloughs, riverbanks, wet pastures, and the edges of pools in alluvial or otherwise wet woodlands; primarily in the northern two-thirds of Wisconsin. In northern Wisconsin the species often grows in wet hardwood forests and vernal pools or other wet depressions within otherwise mesic forests; common also in conifer bogs, alder thickets, cedar swamps, marshes, and wet grasslands. Most southern Wisconsin populations are concentrated along the lower Wisconsin River, with a few additional populations along Lake Michigan. Typical floodplain forest associates include silver maple, American elm, buttonbush (*Cephalanthus occidentalis*), *Carex grayi, C. lupulina, C. muskingumensis, C. tribuloides, C. typhina,* moneywort (*Lysimachia nummularia*), water-parsnip (*Sium suave*). Additional associates in northern forests include *C. brunnescens, C. gracillima, C. intumescens, C. projecta.*

Similar species. Tuckerman's sedge is one of Wisconsin's most easily recognized species: the combination of *inflated, distinctly veined perigynia ≥ 4.5 mm wide* and *awnless pistillate scales with entire margins* suffices to distinguish the species. A *dimple (indentation) on the achene angle* clinches the identification, but identification usually does not require dissecting the perigynium.

Plant and perigynium of *Carex tuckermanii*. The terminal spike in this individual is androgynous, bearing a single perigynium at the base.

84. CAREX UTRICULATA—COMMON YELLOW LAKE SEDGE
section Vesicariae
[= Carex rostrata var. utriculata]; Latin: leather bag or bottle, as in "utricle."

Spongy bases, cylindrical pistillate spikes, and small, glabrous, distinctly beaked, inflated perigynia distinguish this sedge from most others in Wisconsin's open wetlands. Flowers May to July, fruits June to July, perigynia falling in August and September.

Plants often forming large colonies; rhizomes elongate; *bases spongy-thickened*, brown or red-tinged. **Culms** 0.3–1 m tall, *smooth or moderately roughened on the angles beneath the inflorescence.* **Leaf blades** mostly 5–12 mm wide, often yellow, the *wider blades distinctly septate-nodulose.* **Upper 2–5 spikes** staminate, occasionally gynecandrous, tipped with perigynia. **Lower spikes** pistillate, occasionally androgynous, tipped with male flowers, ascending or nodding, oblong to cylindrical, *the longest up to 7–10 cm long.* **Pistillate scales** awnless, margins entire. **Perigynia** inflated, 4–8.5 mm long, strongly veined, glossy; beaks distinct, short-toothed.

Habitat and state range. A variety of open wetlands, including fens, bogs, sedge meadows, wet ditches, marshes, lake margins, and floodplains, often sites that are inundated for a good portion of the growing season throughout Wisconsin, more common in central Wisconsin and the northern half of the state. Common associates include *Carex lacustris, C. stricta,* marsh milkweed (*Asclepias incarnata*), bluejoint grass (*Calamagrostis canadensis*), marsh bell-flower (*Campanula aparinoides*), water-hemlock (*Cicuta bulbifera*), pond sedge (*Dulichium arundinaceum*), common spike-rush (*Eleocharis palustris*), stiff bedstraw (*Galium tinctorium*), manna grasses (*Glyceria* spp.), sensitive fern (*Onoclea sensibilis*), wool-grass (*Scirpus cyperinus*), marsh skullcap (*Scutelleria galericulata*), blue skullcap (*S. lateriflora*), cattails (*Typha* spp.).

Similar species. Common yellow lake sedge is often treated as a variety of *C. rostrata* (beaked sedge [85]), a species of boggy wetlands in far northern Wisconsin that differs in having *leaf blades U-shaped in cross-section, papillose on the upper (adaxial) surface (20× magnification), generally < 5 mm wide;* and *perigynia mostly < 6 mm long.* Most reports of *C. rostrata* in North America are actually *C. utriculata.* **Carex vesicaria** (blister sedge [83]) differs from *C. rostrata* and *C. utriculata* in being *cespitose with short rhizomes, the basal leaf sheaths not spongy,* and *culms roughened on the angles beneath the inflorescence* (the latter character is difficult to interpret in many specimens). **Carex lacustris** (lake sedge [86]) bears a superficial vegetative resemblance to *C. utriculata* but has *reddish, pinnate-fibrillose basal leaf sheaths* and *very short perigynium beaks tipped with 2–3 short teeth.*

Plant, pistillate spike, and perigynium of *Carex utriculata*. Perigynium and pistillate spike dimensions can vary approximately twofold from one plant to another; as a consequence, the species often looks very different at different sites.

86. *CAREX LACUSTRIS*—LAKE SEDGE, RIP-GUT SEDGE
section *Paludosae*
[= *Carex riparia* var. *lacustris*]; Latin: lake, as in "lacustrine."

The stiff bluish shoots of this species emerge in marshes and sedge meadows in March and April, one of spring's first signs. Flowers May to June, fruits May to July, perigynia typically falling by the beginning of August.

Plants *forming large, mostly vegetative colonies;* rhizomes ropelike, frequently > 0.5 cm thick; bases *thick, spongy, clothed in sheaths of the previous year's leaves,* the current year's leaf sheaths *red, ladder-fibrillose.* Culms ≤ 1.3 m tall, harshly scabrous. Leaf blades W-shaped in cross-section, 8–20 mm broad, often septate-nodulose, the bumps in the interveinal areas of wider leaves reminiscent of a player-piano roll. Lower bracts roughly equaling the tip of the inflorescence. Upper 3–5 spikes staminate. Lower spikes pistillate, occasionally androgynous, tipped with staminate flowers, erect, occasionally drooping, 1.5–10 cm long. Perigynia 5–7 mm long, the poorly developed beak bearing *two prominent teeth and often a third shorter tooth.*

Habitat and state range. Most common in sedge meadows and marsh edges, occurring in almost any wetland with shallow standing water. F. W. Hamerstrom reports that "muskrats eat root-stocks, bases of culms and toward young tips" (1937, herbarium label). Common associates include *Carex stricta, C. pellita, C. utriculata,* sensitive fern (*Onoclea sensibilis*), marsh fern (*Thelypteris palustris*), marsh milkweed (*Asclepias incarnata*), bluejoint grass (*Calamagrostis canadensis*), marsh bellflower (*Campanula aparinoides*), spring-cress (*Cardamine bulbosa*), water-hemlock (*Cicuta maculata*), joe-pye weed (*Eupatorium maculatum*), boneset (*Eupatorium perfoliatum*), grass-leaved goldenrod (*Euthamia graminifolia*), tear-thumb (*Polygonum sagittatum*), wool-grass (*Scirpus cyperinus*), marsh skullcap (*Scutellaria galericulata*), common bur reed (*Sparganium eurycarpum*), cattails (*Typha* spp.).

Similar species. Perigynia of *Carex lacustris* are unlikely to be mistaken for anything else in our flora. Vegetatively, the red, ladder-fibrillose basal leaf sheaths and bluish foliage recall C. *stricta* (tussock sedge [70]), which differs in having *narrower leaf blades (4–6 mm wide)* and *plant bases not thickened or spongy.* Lake sedge in vegetative condition is also sometimes confused with C. *utriculata* (common yellow lake sedge [84]), which typically has *yellow to yellow-green foliage* and *leaf blades flat or V-shaped in cross-section, 5–12 mm wide.*

Plant, pistillate spike, and perigynium of *Carex lacustris*. Even perigynia that lack the characteristic diminutive third beak tooth are easily recognized by their distinctive shape.

88. *CAREX PELLITA*—BROAD-LEAVED WOOLLY SEDGE

section *Paludosae*

[= *Carex lanuginosa*]; Latin: skin, leather. The obsolete English word "pell" refers to the skin, or "velvet," that covers a deer's antlers, a lovely analogy for the pubescence on the perigynium of this species and its relatives.

The densely flowered spikes of tiny, peachlike perigynia in this species and its close relative, *C. lasiocarpa*, cannot be mistaken for anything else in the flora. Flowers May to June, fruits June, perigynia falling July and August.

Plants colonial, producing shoots singly; rhizomes elongate; basal leaf sheaths reddish, ladder-fibrillose. **Culms** 15–90 cm tall. **Leaf blades** *flat or* M-*shaped in cross-section*, 2–6 mm wide, tapering to a long-pointed apex. **Upper 1–3 spikes** staminate, erect, elevated above the pistillate spikes. **Lower 1–4 spikes** pistillate, oblong-cylindrical, densely packed with perigynia. **Perigynia** 2.5–5 mm long, *densely pubescent;* beak ≤ 1.5 mm long, teeth sharp, 0.5–1 mm long.

Habitat and state range. Common in wet, sunny, or sparsely wooded areas throughout most of the state, especially in sandy or disturbed soils; more common in the southern half of the state, absent from the northern highlands. *Carex pellita* forms colonies in wet prairies, low fields, sedge meadows, and marshes. Vegetative shoots often spread by rhizomes to adjacent roadsides, railroad embankments, or trails. Typical associates include *C. bebbii, C. hystericina, C. tenera, C. tetanica, C. vulpinoidea,* and other wetland species.

Similar species. *Carex pellita* is most similar to **C. lasiocarpa** (**wiregrass** [89]), a species of bogs, marshes, very wet sedge meadows, and other open wetlands throughout the glaciated portions of the state, more common than *C. pellita* in northern Wisconsin. Both species are strongly rhizomatous, but *C. lasiocarpa* frequently *forms floating mats in standing water or sphagnum bogs* and has *involute leaf blade margins*. **Carex houghtoniana** (**Houghton's sedge** [87]) inhabits dry to mesic sandy and rocky soils, rarely if ever as wet as those of *C. pellita* or *C. lasiocarpa,* predominantly in the northern third of the state. It is distinguished by its *longer perigynia (4.5–6.5 mm)* with *veins visible beneath the pubescence.* The unrelated **C. stricta** (**tussock sedge** [70]) has red, pinnate-fibrillose basal leaf sheaths similar to those of *C. lasiocarpa* and *C. pellita,* but intact leaf sheaths of the latter two species are frequently tinged yellowish brown, unlike those of *C. stricta.*

Plant, inflorescence, and perigynium of *Carex pellita*. The very similar *C. lasiocarpa* is distinguishable based on leaf morphology and habitat (as discussed above); otherwise, the two plants are nearly identical.

90. CAREX TRICHOCARPA—HAIRY-FRUIT LAKE SEDGE

section *Carex*

Greek: hair (as in "trichome") and fruit (as in "carpel").

All species of this section form tall, slender, vegetative culms that are generally far more numerous than the fertile culms. Leaf sheath characteristics are so distinctive that even sterile specimens are often readily identifiable. Flowers May to June, fruits June to July.

Plants *forming large colonies that are generally composed predominantly of vegetative shoots; rhizomes elongate; basal leaf sheaths ladder-fibrillose.* **Vegetative culms** 0.6–1.6 m tall. **Fertile culms** 0.6–1.3 m tall. **Leaf blades** ≤ 9 mm wide, glabrous the wider blades **W**-shaped in cross-section, margins harshly scabrous. **Inner band of the leaf sheaths** glabrous, *tinged reddish to purple at the summit,* this coloration typically forming an elongate, inverted triangle. **Upper 1–8 spikes** staminate. **Lower spikes** pistillate, erect to arching. **Perigynia** 6–11 mm long, *pubescent;* beaks prominent, teeth > 1 mm long.

Habitat and state range. See facing page.

Similar species. Two other species of the section are common in Wisconsin. ***Carex laeviconica*** (**long-toothed lake sedge [93]**) is a floodplain species that ranges to shadier habitats than the other members of the section. Identify it by the combination of *glabrous perigynia* and *glabrous, prominently veined leaf sheaths, the inner band becoming fibrillose.* **Carex atherodes** (**slough sedge [92]**) is most common in standing water < 1 m deep within sedge meadows, marshes, willow swamps, and tamarack bogs. Identify it by the combination of *glabrous perigynia, the pubescence, if any, limited to the beak; leaf sheaths typically pubescent, at least at the summit,* although sheaths and leaves may be glabrous, especially on individuals standing in water; and *leaves papillose and typically long-pubescent on the underside, glabrous on the upper surface.* Leaves of *C. laeviconica* are smooth and thus distinguishable from the odd slough sedge with glabrous leaf sheaths. Allison Dibble (pers. comm., 2007) notes that colonies of *C. atherodes* turn an easily spotted, rich golden green color in October.

A fourth species, **Carex hirta** (**hammer sedge [91]**), is native to Eurasia and known from two highly disturbed sites in Wood and Grant counties, its westernmost reported localities in North America. *Carex hirta* has *perigynia pubescent on all surfaces; pistillate scales glabrous or—unlike any other Wisconsin members of the section—spreading-pubescent;* and *leaf sheaths and blades pubescent, not papillose.*

Habitats of Wisconsin's section *Carex* species

Carex trichocarpa. Generally in floodplain sedge meadows, wet prairies, and marsh edges, ranging occasionally to lowland forests; almost restricted to south of the Tension Zone, with a few outlying populations in Brown, Manitowoc, Ashland, and Bayfield counties. *Carex trichocarpa* is the host for caterpillars of the two-spotted skipper (*Euphyes bimacula*). Associates include *C. pellita, C. stipata, C. stricta*, great angelica (*Angelica atropurpurea*), bristly aster (*Aster puniceus*), blunt-leaf bedstraw (*Galium obtusum*), downy phlox (*Phlox pilosa*), stinging nettle (*Urtica dioica*), golden alexanders (*Zizia aurea*).

Carex atherodes. Wet prairies, sedge meadows, cattail marshes, lakeshores, and bogs; most common in standing water < 1 m deep, ranging into wet, typically unshaded soil. Primarily in the eastern half and northwest quarter of the state. Common associates include willows (*Salix* spp.), *C. aquatilis, C. lacustris, C. stricta, C. sartwellii*, bluejoint grass (*Calamagrostis canadensis*), cattails (*Typha* spp.), and other sedge meadow and marsh species.

Carex laeviconica. Open and wooded alluvial wetlands, slightly more frequently in the former than the latter; uncommon in Wisconsin, where it is at the northeast edge of its range. Primarily along the Mississippi and lower Wisconsin rivers. Hujik (1995) considers this plant, which is rare in Wisconsin and not modal in any of Curtis's (1959) communities, to be a lowland savanna species. Associates include silver maple, buttonbush (*Cephalanthus occidentalis*), arrowheads (*Sagittaria* spp.).

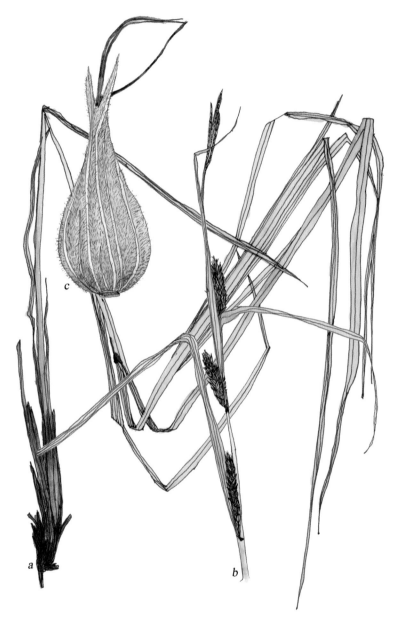

(a) Vegetative shoot, (b) inflorescence, and (c) perigynium of *Carex trichocarpa*. Note the coloration at the apex of the inner band of the leaf sheaths. This coloration is distinctive but variable, ranging from a narrow band to an elongate, inverted triangle that may extend for several centimeters down the sheath. Inner band of the leaf sheath on (d) *C. laeviconica* and (e) *C. atherodes*. Pubescence is not an infallible character for *C. atherodes*, but glabrous individuals can be separated from *C. laeviconica* by the presence of papillae on the undersides of the leaves in the former.

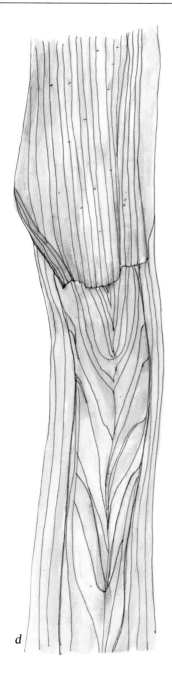

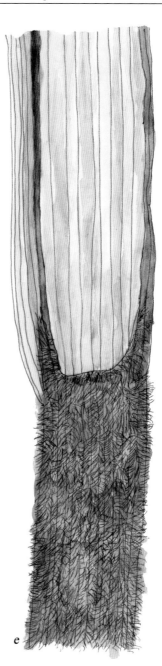

96. CAREX INTUMESCENS—SHINING BUR SEDGE

section *Lupulinae*

Latin: becoming swollen.

Showy spikes of ascending, inflated, teardrop-shaped perigynia characterize this species. Flowers May to June, fruits June to July.

Plants cespitose; rhizomes short; basal leaf sheaths red. **Culms** 0.2–1 m tall. **Sheath of the uppermost leaf** usually ≤ 1.5 cm long. **Leaf blades** mostly 4–8 mm broad; ligule apex rounded. **Bracts** foliose, overtopping the inflorescence. **Terminal spike** staminate, often borne on a slender stalk approximately the same length as the spike. **Lateral spikes** pistillate, 1–4, congested at the tip of the culm, slightly longer than thick, 1–3 cm long. **Perigynia** 2–12 per spike, mostly ascending, narrowly teardrop-shaped, 10–16 mm long, 2.5–6.5 mm wide, *widest below the middle, strongly veined, lustrous;* beak teeth approximately 1 mm long.

Habitat and state range. See facing page.

Similar species. Shining bur sedge superficially resembles three other species in section *Lupulinae.* *Carex grayi* (**Gray's sedge** [97]) has *perigynia radiating outward in all directions, reminiscent of the head of a mace, the surfaces dull or pubescent.* Individuals of *C. grayi* with fewer perigynia may resemble *C. intumescens,* while *C. intumescens* individuals with strongly divergent lower perigynia may recall a depauperate *C. grayi.* Use perigynium surface texture to confirm identifications of the species. *Carex lupulina* (**hop sedge** [98]) has *uppermost leaf sheaths generally > 1.5 cm long; pistillate spikes usually conspicuously longer than wide with up to 80 perigynia;* and *styles bent* (dissect the perigynium to see this). *Carex lupuliformis* (**knobbed hop sedge** [99]), uncommon in the state, suggests a very robust *C. lupulina* with distinct knobs on the achene angles.

The two species of section *Rostrales*—*Carex folliculata* (**long sedge** [94]) and *C. michauxiana* (**Michaux's sedge** [95])—bear a passing resemblance to *C. intumescens* but are very easily distinguished, as both have *brown basal leaf sheaths* and *perigynia barely inflated, > 4 times as long as wide, tapering to the base and apex.*

Habitats of Wisconsin's section *Lupulinae* and *Rostrales* species

Carex intumescens. Common in mesic to wet forests throughout the northern two-thirds of Wisconsin, ranging to all but the southernmost tier of counties, occasional in wet ditches and fields. Typical associates include *C. brunnescens, C. lupulina, C. radiata, C. tuckermanii,* spinulose wood fern (*Dryopteris carthusiana*), drooping wood-reed (*Cinna latifolia*), bunchberry dogwood (*Cornus canadensis*), fowl manna grass (*Glyceria striata*), interrupted fern (*Osmunda claytoniana*), fringed bindweed (*Polygonum cilinode*), American starflower (*Trientalis borealis*).

Carex lupulina. Most common in alluvial forests, especially along the edges of sloughs; also found in alluvial wet meadows, marshes, ditches, lakeshores, and wet depressions in otherwise mesic forests throughout the state. Common associates include black ash, silver maple, river birch, *C. crinita, C. muskingumensis, C. projecta, C. tribuloides, C. tuckermanii, C. typhina,* virgin's-bower (*Clematis virginiana*), Virginia wild-rye (*Elymus virginicus*), fowl manna grass (*Glyceria striata*), moneywort (*Lysimachia nummularia*), sensitive fern (*Onoclea sensibilis*).

Carex grayi. Alluvial forests, especially along the edges of sloughs; along larger rivers, primarily in southern Wisconsin. Common associates include silver maple, sugar maple, river birch, green ash, *C. lupulina, C. tribuloides,* green dragon (*Arisaema dracontium*), jack-in-the-pulpit (*A. triphyllum*), carrion-flowers (*Smilax* spp.).

Carex folliculata. Locally abundant in wet forests, boggy thickets, and sandy wetlands, centered on Jackson County. Common associates include red maple, winterberry, alder, *C. debilis, C. echinata, C. gynandra, C. intumescens,* blue-bead-lily (*Clintonia borealis*), Canadian rush (*Juncus canadensis*), cinnamon fern (*Osmunda cinnamomea*), skunk-cabbage (*Symplocarpus foetidus*), bog fern (*Thelypteris simulata*).

Carex michauxiana. Locally abundant in bogs, bottomland forests, and wet acidic soils of the Bayfield area. Associates include black spruce, larch, white cedar, *C. exilis, C. lasiocarpa, C. limosa, C. livida,* swamp-pink (*Arethusa bulbosa*), white beak-rush (*Rhynchospora alba*), and other northern bog species.

(a) Plant, (b) inflorescence, and (c) perigynium of *Carex intumescens*. (d) Pistillate spike of *C. grayi* and (e) inflorescence of *C. lupulina*.

Common Wisconsin Species
of *Carex* Subgenus *Vignea*

101. *Carex gynocrates*—Northern bog sedge

section *Physoglochin*

Greek: female and ruling, presumably referring to the distinctive pistillate spike on female plants.

The solitary pistillate spikes with plump, dark, divergent perigynia are highly distinctive. Flowers May, fruits June to July.

Plants clonal, shoots arising singly or in dense tufts; *rhizomes threadlike, < 1 mm thick.* **Culms** 0.1–0.3 m tall. **Leaf blades** < 1 mm wide, wiry. **Inflorescence** unispicate, 5–15 mm long, usually pistillate, less often staminate or androgynous with 1–2 perigynia at the base; bracts lacking. **Staminate spike** *approximately twice the thickness of the culm.* **Pistillate spike** 5–15 mm long, approximately twice as long as wide. **Perigynia** generally 5–15 per spike, *oriented perpendicular to the culm at maturity* (divergent) or slightly reflexed, convex on both faces, slightly flattened in cross-section, dark brown at maturity, 3–3.5 mm long, 15–20-veined (often distinctly so), occasionally 2-ribbed; beak 0.5 mm long.

Habitat and state range. Sphagnum bogs, white cedar swamps, and calcareous swales adjacent to Lake Michigan; primarily in the northeastern quarter of the state and Door County, with disjunct populations in Taylor and Ozaukee counties. Associates include black spruce, larch, white cedar, *Carex tenuiflora, C. crawei,* white beak-rush (*Rhynchospora alba*), small round-leaved orchis (*Amerorchis rotundifolia*), fringed orchids (*Platanthera* spp.), swamp saxifrage (*Saxifraga pensylvanica*), arrow-grass (*Triglochin maritima*).

Similar species. The other unispicate sedges that occur in similar habitats are **Carex pauciflora** (few-flowered bog sedge [3]), *C. leptalea* (bristle-stalked sedge [4]), and *C. exilis* (coastal star sedge [131]). None has filiform leaf blades. Leaf blades of *C. exilis* are 0.5–1.5 mm wide with involute margins, but the widest are generally > 1 mm wide; *C. exilis* also has short, inconspicuous rhizomes. *Carex gynocrates* is considered by some authors to range to East Asia (Siberia), but other authors consider the Siberian populations a separate species.

Plant and pistillate spike of *Carex gynocrates*. The plump, divergent perigynia make the plant unmistakable in our flora.

103. *CAREX SICCATA*—RUNNING PRAIRIE SEDGE

section *Ammoglochin*

The name *Carex foenea* is misapplied to this species in many treatments. The true *C. foenea* [141] is in section *Ovales*. Latin: dry.

The unkempt appearance of the inflorescence and the long-creeping rhizomes make *Carex siccata* distinct in its favored sandy locales. Flowers May, fruits June to July.

Plants *colonial, shoots arising singly; rhizomes brown, scaly, long-creeping.* **Culms** 0.1–1 m tall, roughened on the angles. **Leaves** stiff, the inner band of the sheath hyaline at least at the summit, the blade 1–3 (occasionally 4) mm wide. **Spikes** *pistillate, staminate, gynecandrous,* or *androgynous,* a mix of these common in a single inflorescence, overlapping or the lower distinct. **Pistillate scales** reddish brown, shorter than to equaling the perigynium, *often covering much of the perigynium body.* **Perigynia** narrowly winged, veinless or few-veined on the inner face, strongly veined on the back, variable in shape and venation.

Habitat and state range. Most common in dry sand prairies, black oak and jack pine barrens, and sandy woods, ranging to occasional wet prairies; mostly in the southern half of the state but ranging as far north as Douglas and Marinette counties. Typical associates include *Carex muehlenbergii,* pussytoes (*Antennaria* spp.), wormwood (*Artemisia campestris*), butterfly milkweed (*Asclepias tuberosa*), bastard toadflax (*Comandra umbellata*), sweet-fern (*Comptonia peregrina*), prairie coreopsis (*Coreopsis palmata*), frostweeds (*Helianthemum* spp.), round-headed bush-clover (*Lespedeza capitata*), puccoons (*Lithospermum* spp.), lupine (*Lupinus perennis*), porcupine grass (*Stipa spartea*).

Similar species. Running prairie sedge is occasionally confused with two species. *Carex sartwellii* (**Running marsh sedge** [102]) generally grows in sedge meadows and wet prairies and thus usually does not overlap in habitat with *C. siccata. Carex sartwellii* forms *tall vegetative culms, the inner band of the leaf sheaths green and distinctly veined.* Like *C. siccata,* this species reproduces by rhizomes and often forms large clones dominated by vegetative shoots. The more distantly related *C. praegracilis* (**freeway sedge** [100]) is a long-rhizomatous, western North American species that has spread rapidly eastward since 1900. Freeway sedge has been collected along a few roadsides in Wisconsin. Distinguish it from both of the "running sedges" by the fact that freeway sedge generally has *unisexual inflorescences.* The species does not grow in wetlands. It could grow with *C. siccata* in dry disturbed soils.

(a) Plant, (b) inflorescence, and (c) perigynium of *Carex siccata*. Note the elongate rhizome. (d) Leaf sheath of *C. sartwellii*, showing the distinctive venation of the inner band.

104. *Carex vulpinoidea*—Fox sedge
section *Multiflorae*
Latin: fox.

One of Wisconsin's most common wetland sedges, easily recognized by its straight, narrow, compound inflorescence with numerous needlelike bracts as well as the tight leaf sheaths that are corrugated on the inner band. Flowers May to June, fruits June to July, perigynia mostly falling by September.

Plants strongly cespitose. **Culms** *firm, not spongy or easily crushed*, scabrous, the margins unwinged, 0.1–1 m tall, 2 mm thick. **Leaf sheaths** tight, *the inner band corrugated.* **Leaf blades** *longer than the culms*, ≤ 5 mm wide; ligules rounded or notched at the apex. **Inflorescence** compound, often cylindrical, 3–10 cm long; bracts setaceous. **Spikes** androgynous, at least the *lowest branching* and often separate. **Perigynia** green to brown at maturity, *tapering gradually to the beak*, 2–3 mm long, veinless on the inner face; *beak one-third to half the total perigynium length.*

Habitat and state range. Common in marshes, wet forest edges, alluvial woods, lake and stream edges; occasional in wet prairies, fens, and white cedar swamps; infrequent in bogs and standing water throughout the state. Fox sedge tolerates disturbance. Associates include *Carex bebbii, C. hystericina, C. interior, C. pellita, C. stipata,* marsh milkweed (*Asclepias incarnata*), Dudley's rush (*Juncus dudleyi*), sensitive fern (*Onoclea sensibilis*), and other species of open to moderately wooded and frequently calcareous wetlands.

Similar species. Fox sedge closely resembles **Carex annectens** (yellow-headed fox sedge [105]), a species of wet to mesic prairies and meadows and other open sandy areas, primarily in the western half of Wisconsin. *Carex annectens* is most easily distinguished by its *leaf blades, which are shorter than the culms,* and by the *yellow coloration of its ripe perigynia, which narrow abruptly to a beak that is ≤ one-quarter the total perigynium length.* Fox sedge is sometimes confused with *C. stipata* (owl-fruit sedge [108]), which also has a compound, cylindrical inflorescence and leaf sheaths corrugated on the inner faces, but the *thick, spongy culms* of *C. stipata* readily distinguish it from *C. vulpinoidea* and *C. annectens.*

(a) Plant, (b) lower spike, and (c) perigynium of *Carex vulpinoidea*. (d) Perigynium of *C. annectens*. Note the relatively long leaves of *C. vulpinoidea;* the difference in perigynium shape and beak length between *C. vulpinoidea* and *C. annectens;* and the compound lower spike, which is typical of both species.

106. *Carex diandra*—Bog panicled sedge
section *Heleoglochin*
Greek: 2 anthers.

This tussock-forming wetland species is distinguished by the red dots on the inner band of its leaf sheaths and by its tiny perigynia, which become dark chestnut brown at maturity. Flowers May to June, fruits June, perigynia falling in July or August.

Plants cespitose, often forming small tussocks; rhizomes short; basal leaf sheaths bladeless, reddish brown. Culms 0.1–1 m tall. Inner band of the leaf sheaths *whitish-hyaline, red-dotted at least near the summit.* Leaf blades frequently equaling the tip of the inflorescence, 1–2.5 mm wide, long-tapering to a narrow apex; ligules ≥ 8 mm long. Inflorescence compound, stiff or arching, 2–6 cm long, 0.5–1.5 cm thick; bracts few, setaceous, short. Spikes androgynous, the lower branched. Pistillate scales roughly as wide as the perigynium near the base. Perigynia dark and glossy at maturity, the inner face strongly convex, veinless, the *back conspicuously 2-ribbed,* approximately 2.5 (occasionally 3) mm long, *base spongy-swollen;* beak triangular, approximately 1 mm long.

Habitat and state range. Most common in bogs, fens, white cedar swamps, peaty sedge meadows, and other peaty wetlands; occasional on sandy riverbanks and lakeshores; primarily in the eastern half of Wisconsin and wetlands along Lake Superior, with scattered populations in northwestern Wisconsin. Common associates include white cedar, tamarack, black spruce, *Carex aquatilis, C. comosa, C. lasiocarpa, C. leptalea, C. stricta,* marsh fern (*Thelypteris palustris*), marsh bellflower (*Campanula aparinoides*), bogbean (*Menyanthes trifoliata*), pitcher-plant (*Sarracenia purpurea*).

Similar species. *Carex diandra* resembles *C. prairea* (prairie sedge [107]), a species of fens, sedge meadows, occasional wet prairies, and other calcareous wetlands, primarily in the eastern half of the state. Distinguish prairie sedge by the *copper coloring at the summit of the inner band of its leaf sheaths.* Moreover, *perigynia of C.* prairea *are nearly concealed by their pistillate scales and are flat on the inner face,* while those of *C. diandra* are exposed and convex on both faces. *Carex sartwellii* (running marsh sedge [102]) is similar in range and habitat to prairie sedge and has inflorescences that bear a superficial resemblance, but *C. sartwellii has spikes simple rather than compound* and *the inner band of the leaf sheaths is green and strongly veined.*

(a) Sheath, (b) inflorescence, and (c) perigynium of *Carex prairea*. (d) Sheath, (e) inflorescence, and (f) perigynium of *C. diandra*.

108. *CAREX STIPATA*—OWL-FRUIT SEDGE

section *Vulpinae*

Latin: crowded.

The thick, bristly inflorescence, sharply winged culms that are easily crushed between the fingers, and corrugated inner band of the leaf sheaths make this one of North America's most recognizable wetland sedges. Vegetative characters often suffice to identify the species. Flowers mid- to late May, fruits June, perigynia generally falling in July or August.

Plants cespitose; basal leaf sheaths pale to brown. **Culms** to > 1 m tall and 7 mm thick, *spongy, easily compressed,* the *margins narrowly winged, scabrous.* **Inner band of the leaf sheaths** *corrugated,* whitish-hyaline, firm but not thickened at the summit, unpigmented. **Leaf blades** to 1 (occasionally 1.5) cm wide. **Inflorescence** *compound, at least the lower spikes branched,* generally with at least a few needlelike bracts, *bristly with perigynium beaks and bracts,* 5–15 cm long. **Spikes** androgynous. **Perigynia** 4–6 mm long, both faces veined, *base spongy;* beak elongate, approximately half the length of the entire perigynium, margins scabrous.

Habitat and state range. Common in a variety of moist, often shaded habitats such as river or pond margins, sedge meadows, and wet roadsides; tolerant of disturbance; throughout Wisconsin. Typical associates include *Carex pellita, C. trichocarpa,* wild garlic (*Allium canadense*), bristly buttercup (*Ranunculus hispidus*), tall meadow-rue (*Thalictrum dasycarpum*), and a wide range of wetland generalists.

Similar species. Three other species from this section are found in Wisconsin, though none as commonly as *Carex stipata.* **Carex alopecoidea** (foxtail sedge [111]) is found primarily in floodplains of the southern two tiers of Wisconsin counties. Distinguish foxtail sedge from owl-fruit sedge by the *shorter (≤ 4 mm long), veinless perigynia that are not spongy at the base* and the *smooth, red- or purple-spotted inner band of the leaf sheaths.* **Carex laevivaginata** (smooth-sheathed sedge [110]) is a state endangered species very similar to *C. stipata* in overall appearance but distinguished by an obvious *thickening at the mouth of the inner band of the leaf sheath,* which is otherwise smooth. **Carex crus-corvi** (crow-foot fox sedge [109]) is a state endangered species known from ditches and wet hollows in Milwaukee and Waukesha counties. It has a *long-branched inflorescence* and *perigynia with slender, parallel-sided beaks that are roughly three times as long as the body.*

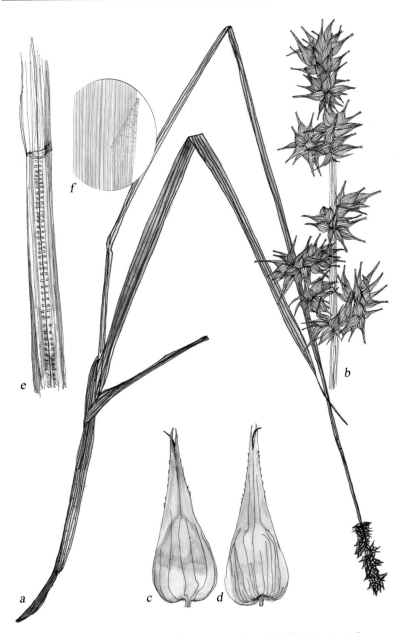

(a) Plant, (b) inflorescence, (c) perigynium inner face, (d) perigynium outer face, and (e) leaf sheath inner band of *Carex stipata*. Compare the corrugated inner band of the leaf sheath in *C. stipata* with (f) the smooth, dotted inner band of the leaf sheath in *C. alopecoidea*.

112. *Carex chordorrhiza*—Creeping sedge
section *Chordorrhizae*
Latin and Greek: string and root.

This bog species is easily recognized by its elongated stolons. The capitate inflorescence and perigynia with rounded margins and abrupt, snoutlike beaks are distinctive. Flowers May to late June, fruits June to July.

Plants *prominently stoloniferous, most shoots arising from nodes of the previous season's vegetative shoots, which extend to 1.2 m long;* rhizomes short. **Culms** 5–40 cm tall, bearing a few short leaves. **Leaf blades** ≤ 3 mm wide. **Inflorescence** capitate, somewhat longer than wide; bracts short, inconspicuous. **Spikes** 2–5, androgynous, *densely congested.* **Perigynia** rounded on both faces and on the margins, dark brown at maturity, 2–3.5 mm long, 1.5–2 mm wide, strongly veined, glossy, base short-stipitate; beak < 0.5 mm, the apex bidentate or irregularly cleft.

Habitat and state range. Sphagnum bogs throughout the state; occasional in other peaty wetlands such as fens, white cedar swamps, northern sedge meadows. Common associates include larch, black spruce, poison sumac, *Carex disperma, C. lasiocarpa, C. limosa, C. livida, C. oligosperma, C. trisperma,* bog-rosemary (*Andromeda glaucophylla*), leather-leaf (*Chamaedaphne calyculata*), huckleberry (*Gaylussacia baccata*), Labrador-tea (*Ledum groenlandicum*), pitcher-plant (*Sarracenia purpurea*), cranberry (*Vaccinium macrocarpon*), and other typical bog species.

Similar species. *Carex chordorrhiza* may be overlooked but not easily misidentified.

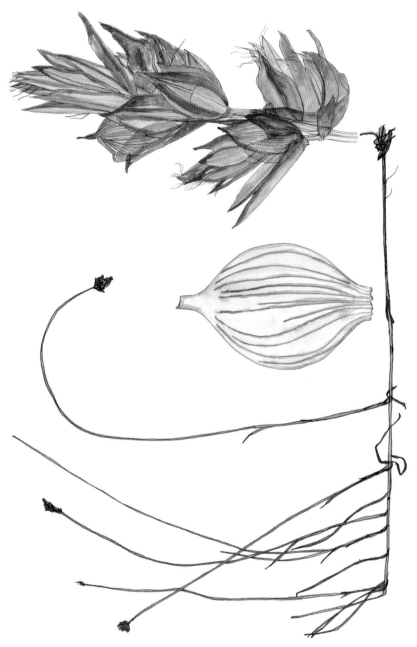

Plant, inflorescence, and perigynium of *Carex chordorrhiza.*

114. *CAREX CEPHALOIDEA*—CLUSTERED BRACTED SEDGE
section *Phaestoglochin*
Greek: head-shaped.

The loose leaf sheaths, dense inflorescence of androgynous spikes, and relatively short pistillate scales distinguish this species from most others in our flora. Flowers May, fruits June, perigynia falling in July. **Plants** densely cespitose. **Culms** 0.3–1.3 m tall, 2.5–4 mm thick at the base but not spongy, narrower above. **Leaf sheaths** *enclosing the culm loosely, back whitened between the veins with prominent cross-veins, inner band frequently corrugated.* **Leaf blades** 4–8 mm wide. **Inflorescence** 1.5 cm long, 1 cm wide, lowest inflorescence internode < twice as long as the lowest spike. **Spikes** androgynous, overlapping or the lowest separate. **Pistillate scales** concealing ≤ half of the perigynium, apex acute to acuminate, or short awned. **Perigynia** green, 3–4.5 mm long, margins acute to narrowly winged.

Habitat and state range. Mostly in mesic forests, often disturbed, but also found at woodland edges; primarily in the southern half of the state, scattered in the north. Associates include *Carex blanda, C. bromoides, C. deweyana, C. echinodes, C. hirtifolia, C. normalis, C. radiata, C. rosea,* and common mesic forest species such as Canadian honewort (*Cryptotaenia canadensis*), cleavers (*Galium aparine*), wild geranium (*Geranium maculatum*), Virginia waterleaf (*Hydrophyllum virginianum*).

Similar species. Superficially this species resembles *Carex normalis* (**greater straw sedge [155]**), which grows in similar habitats but has *tight sheaths* and *gynecandrous spikes. Carex gravida* (**heavy sedge [115]**) is a very similar species that occupies open, disturbed habitats and occasional dry prairies in the southern half of the state, where it is probably more common than *C. cephaloidea. Pistillate scales in* C. gravida *tend to be awned or have a narrow, elongate apex,* and they *generally conceal > half of the perigynium body.* Perigynia in *C. gravida* are spongy at the base, and at maturity they become dark brown and swollen. *Carex aggregata* (**aggregated sedge [116]**) is similar in habitat and appearance to *C. gravida* but known from few collections in southern Wisconsin. Distinguish it by the inner band of the *leaf sheaths, which are yellow or brown and thickened at the summit,* tearing less easily than those of *C. gravida. Carex sparganioides* (**bur-reed sedge [113]**) is a mesic forest species of southern Wisconsin that sometimes approaches *C. cephaloidea* in appearance, but its *open, elongate, often arching inflorescence* makes it the most distinct of these four species.

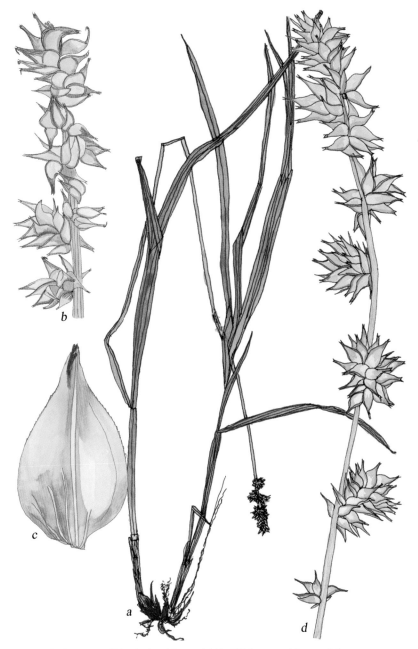

(a) Plant, (b) inflorescence, and (c) perigynium of *Carex cephaloidea*. (d) Inflorescence of *C. sparganioides*.

117. *CAREX ROSEA*—CURLY-STYLED WOOD SEDGE

section *Phaestoglochin*

[= *Carex convoluta*, the name *C. rosea* often misapplied to *C. radiata*]; Latin: pink.

You will recognize the starlike spikes of this species if you have gone sedge hunting in upland forests anywhere in eastern North America. The perigynia with their characteristically tight-curled stigmas drop readily at maturity. Flowers May, fruits June, perigynia falling by late summer.

Plants densely cespitose; rhizomes short. **Culms** 0.2–0.8 m tall, > 1.5 mm wide at the base, 0.5–1 mm wide beneath the inflorescence, *erect or leaning*. **Leaf sheaths** tight. **Leaf blades** 1.8–2.5 mm wide. **Inflorescence** stiff, open, 2–7 cm long. **Spikes** androgynous, few-flowered, the staminate flowers usually inconspicuous at the tips of the spikes. **Perigynia** ascending to divergent or reflexed at maturity, green, 2.5–4 mm long, veinless; base spongy, no more than one-fifth the total perigynium length; beak 0.5–1 mm long, margins serrate; *stigmas thick, tightly coiled.*

Habitat and state range. Common in mesic forests, dry to mesic oak woods, and occasional prairies or wet pastures throughout the state, more common in the southern and eastern halves. Associates include sugar maple, white oak, red oak, shagbark hickory, *Carex cephalophora*, *C. hirtifolia*, *C. pensylvanica*, *C. sprengelii*, *C. woodii*, and such woodland herbs as jack-in-the-pulpit (*Arisaema triphyllum*), may-apple (*Podophyllum peltatum*), small-flowered buttercup (*Ranunculus abortivus*), bristly buttercup (*R. hispidus*).

Similar species. Curly-styled wood sedge is similar to *Carex radiata* (eastern star sedge [118]), which is found in upland forests throughout Wisconsin, usually in slightly moister soils. Typical *C. radiata* is easily distinguished, its *stigmas thinner than those of C.* rosea *and not tightly coiled* and *leaves and culms usually narrower, weaker, falling prostrate by midsummer.* Vegetative shoots of both species could be confused with *C. pensylvanica* (Pennsylvania sedge [15]), but the latter is *long-rhizomatous with reddish, fibrous basal leaf sheaths* and usually an abundance of dead leaves lying on the ground, persistent from the previous year. The starlike spikes of the wetland section *Stellulatae* are superficially similar to those of *C. rosea* and relatives, but *gynecandrous, the base of the terminal spike conspicuously clavate.*

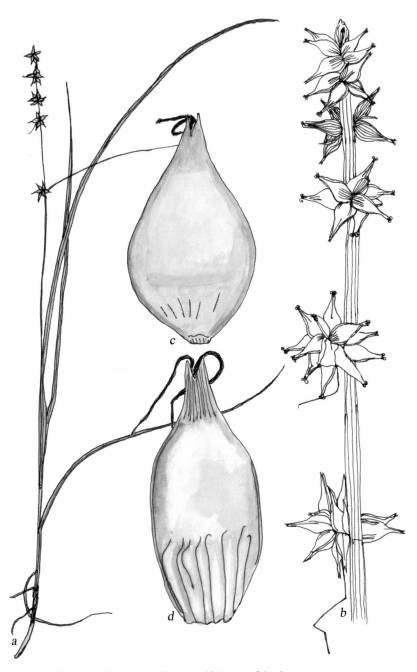

(a) Plant, (b) inflorescence, and (c) perigynium of *Carex rosea*. (d) Perigynium of *C. radiata*.

121. *CAREX CEPHALOPHORA*—OVAL-HEADED SEDGE
section *Phaestoglochin*
Greek: head-bearing.

The compact, headlike inflorescence and stiff, short setaceous bracts of this species are distinctive. Flowers May to June, perigynia usually falling in July. **Plants** cespitose. **Culms** 0.1–1 m tall. **Leaf sheaths** tight, *the inner band smooth, subtly thickened at the summit.* **Leaf blades** 2–5 mm wide. **Inflorescence** capitate; *bracts few to > 10, setaceous.* **Spikes** androgynous. **Pistillate scales** 1–2 mm long, body ≤ two-thirds as long as the perigynia, apex acuminate to short-awned. **Perigynia** pale green to yellow, 2.5–3 mm long, *veinless or weakly veined on the back.*

Habitat and state range. Dry-mesic to mesic deciduous forests, raning to savannas, old fields, and dry to mesic prairies; tolerant of disturbance and often growing along trailsides or disturbed microsites; mostly in the southern third of the state, following the Mississippi River north to Polk County, extending north in the eastern third of the state to Oconto County. Typical mesic forest associates include sugar maple, red oak, *Carex blanda*, *C. rosea*, *C. normalis*, hog-peanut (*Amphicarpaea bracteata*), broad-leaf enchanter's nightshade (*Circaea lutetiana*), Virginia waterleaf (*Hydrophyllum virginianum*), may-apple (*Podophyllum peltatum*).

Similar species. The only common look-alike of *Carex cephalophora* is *C. muehlenbergii* (**Muhlenberg's sedge [119]**), which forms *large, sprawling clumps to nearly 1 m tall* in dry sandy habitats in roughly the same geographic range as *C. cephalophora*. Muhlenberg's sedge is distinguished by the *thin, often corrugated inner face of its leaf sheaths, yellowish and thickened at the summit;* coarser, more robust inflorescence; *pistillate scales ≥ two-thirds as long as the perigynia;* and *perigynia 3–4 mm long*, strongly veined on the back. Two other similar species appear rarely in Wisconsin lawns. *Carex leavenworthii* (**Leavenworth's sedge [122]**) is native throughout most of the eastern United States but introduced in Wisconsin and known from few localities in the state. Its *perigynia tend to be smoother on the margins* on the upper half and *widest near the base, its pistillate scales acute to short-toothed at the apex.* *Carex spicata* (**spiked bracted sedge [120]**) is a West Asian species introduced in the United States, *tinged red to purple at the base* with an *elongate inflorescence* and *ligule that is longer than wide.*

(a) Plant, (b) inflorescence, and (c) perigynium of *Carex cephalophora*. (d) Inflorescence and (e) perigynium of *C. muehlenbergii*.

123. CAREX DISPERMA—TWO-SEEDED BOG SEDGE
section *Dispermae*
Greek: 2-seeded.

The lax inflorescence of this plant, with perigynia solitary, paired, or in trip-
lets at the inflorescence nodes and precariously attached when ripe, resembles
a fine string of beads. Flowers May, fruits June to July, perigynia dispersing
in July and August.

   Plants fine, loosely cespitose or shoots arising singly; rhizomes elongate,
slender, branching; basal leaf sheaths brown. **Culms** 0.1–0.6 m tall, arched
or nodding, slender. **Leaf blades** soft, weak, generally shorter than the culm,
1–2 mm wide. **Spikes** androgynous, staminate flowers forming an inconspicuous cone at the tip of the spike. **Perigynia** *1–3 (occasionally 6) per spike,
spreading, brown and shiny at maturity,* 2–3 mm long, abruptly narrowed to
a short beak, *margins and both faces rounded,* strongly veined.

   Habitat and state range. Bogs, white cedar swamps, and other cool, peaty
wetlands primarily in northern Wisconsin. Common associates include tamarack, black spruce, balsam fir, white cedar, *Carex brunnescens, C. intumescens, C. leptalea, C. trisperma,* and other common species of wet northern
forests.

   Similar species. *Carex disperma* is sometimes mistaken
for *C. trisperma* (three-seeded bog sedge [125]), but that
species has *gynecandrous spikes* and a *slender bract that
equals the inflorescence in length.* Immature material may
also resemble depauperate, immature *C. radiata* (eastern
star sedge [118]) or *C. rosea* (curly-styled wood sedge
[117]), but these species and their relatives have *short
rhizomes* and are not found in the boggy habitats that
favor *C. disperma.* At maturity they could not be mistaken
for *C. disperma.*

Inflorescence of *Carex disperma.*

125. *Carex trisperma*—Three-seeded bog sedge
section *Glareosae*
Greek: 3-seeded, presumably referring to the commonly 3-flowered spikes.

The long, bristlelike lowest bract of the inflorescence makes even immature individuals of this species easy to identify. Flowers May to June, fruits June to July, perigynia often persisting through August or September.

Plants loosely cespitose, sometimes appearing stoloniferous, as new shoots often sprout from lower nodes at or below the sphagnum surface; rhizomes elongate, slender. **Culms** 0.1–0.6 m tall, *narrow and often arching.* **Leaf blades** 0.5–2 mm wide. **Lowest bract** *bristlelike, equaling or exceeding the uppermost spike.* **Spikes** 2–3, gynecandrous, compact, staminate flowers usually inconspicuous. **Perigynia** ovoid, 1–5 per spike, tapering to the beak, brown at maturity, 2.5–3.5 mm long, 2-ribbed, *finely but distinctly many-veined.*

Habitat and state range. Cool, sphagnous wetlands, including bogs, white cedar swamps, and peaty hummocks in black ash swamps and boreal forests, predominantly in northern Wisconsin, additional populations in appropriate habitats of central and southeastern Wisconsin. Typical associates nclude *Carex arctata, C. brunnescens, C. canescens, C. crinita, C. echinata, C. limosa, C. oligosperma,* bog-loving ericads such as bog-rosemary (*Andromeda glaucophylla*), leather-leaf (*Chamaedaphne calyculata*), creeping-snowberry (*Gaultheria hispidula*), bog-laurel (*Kalmia polifolia*), Labrador-tea (*Ledum groenlandicum*), and other herbs of cool, wet northern forests, such as blue-bead-lily (*Clintonia borealis*), gold-thread (*Coptis trifolia*), and twin-flower (*Linnaea borealis*).

Similar species. There are two ecologically distinct varieties in *Carex trisperma*: *C. trisperma* var. *billingsii*, which has leaf blades ≤ 0.5 mm wide and 1–2 perigynia per spike; and typical *C. trisperma* var. *trisperma*, with leaf blades to 2 mm wide and 1–5 (often 3–4) perigynia per spike (Kirschbaum 2007). Wisconsin appears only to have the typical variety. Superficially, *C. trisperma* may be confused with the similarly slender and few-flowered species *C. disperma* (**two-seeded bog sedge** [123]), which differs in its *androgynous spikes, plumper perigynia,* and *short bracts.*

Inflorescence of *Carex trisperma.*

127. CAREX CANESCENS—SILVERY SEDGE
section *Glareosae*
Latin: hoary, referring to the grayish green foliage.

The densely flowered spikes of this common plant are reminiscent of tiny pinecones or clusters of ripe achenes on a buttercup. The inflorescence is generally straight, but a variety with long and sometimes arching inflorescences is also recognized (*Carex canescens* ssp. *disjuncta*). Flowers May to June, fruits June to July, perigynia falling soon after ripening, generally in July.

Plants cespitose; rhizomes short. **Culms** erect, 10–90 cm tall. **Leaf blades** erect, *pale green to grayish green*, sometimes nearly equaling the uppermost spike, 2–4 mm wide. **Lowest bract** bristlelike, short or inconspicuous, *usually not as long as the lowest spike*. **Spikes** 4–8, gynecandrous; terminal spike narrowed to a short, clavate, staminate base. **Perigynia** *10–20 per spike*, convex on the back and flat or slightly convex on the inner face, 2–3 mm long, weakly veined on both faces; beak short, shallowly notched at the apex, smooth or serrulate at the base.

Habitat and state range. Primarily in bogs and peaty wetlands in northern Wisconsin and the Baraboo Hills. Common associates include tamarack, black spruce, speckled alder, leather-leaf (*Chamaedaphne calyculata*), sensitive fern (*Onoclea sensibilis*), *Carex aquatilis, C. echinata, C. lasiocarpa, C. leptalea, C. limosa, C. magellanica, C. oligosperma, C. pauciflora, C. trisperma.*

Similar species. Silvery sedge could be confused with **Carex brunnescens** (**green bog sedge** [128]), which generally has *5–10 perigynia per spike* and *leaf blades 0.5–2 mm wide*. **Carex arcta** (**northern clustered sedge** [126]) is a less common species of northern wet forests, often in floodplains, with *spikes many-flowered, overlapping, often indistinguishable*, and *perigynia widest at the base.*

Inflorescence of *Carex canescens.*

128. *Carex brunnescens*—Green bog sedge
section *Glareosae*
Latin: brown, presumably referring to the color of the perigynia at maturity.

This common sedge, reminiscent of a *Carex rosea* with immature perigynia, is easily recognized by its straight or arching inflorescence and few-flowered, gynecandrous spikes widely separated from one another. Flowers May to June, fruits May to July, perigynia generally falling very soon after ripening in July.

Plants cespitose. **Culms** 10–90 cm tall, erect or arching. **Leaf blades** numerous, 1–1.5 mm wide (ssp. *sphaerostachya*) or 1.5–2.5 mm wide (ssp. *brunnescens*). **Lowest bract** bristlelike, *usually not overtopping the inflorescence.* **Spikes** 5–10, gynecandrous, approximately as long as wide to slightly longer, the lower widely separated and the upper overlapping; *terminal spike frequently with a short, clavate, staminate base not quite as long as the pistillate portion.* **Perigynia** *5–10 per spike, spreading,* 2–2.25 mm long, convex on the back and flat or convex on the ventral face, *subtly or indistinctly veined,* margins rounded; beak short, flattened, margins finely serrate.

Habitat and state range. Cool, mesic to wet forests, ranging to white cedar swamps, bog edges, and occasional oak woodlands; common in the northern third of the state and in central Wisconsin, with several additional populations in the counties adjacent to Lake Michigan. Like all members of this section, *Carex brunnescens* favors peaty soils, but it is probably more tolerant of mineral soils than the other *Glareosae.* Typical mesic forest associates include sugar maple, yellow birch, paper birch, basswood, *C. deweyana, C. projecta,* blue-bead-lily (*Clintonia borealis*), fragrant bedstraw (*Galium triflorum*), rough-leaved rice grass (*Oryzopsis asperifolia*). Typical boreal to northern wet-mesic forest associates include white spruce, balsam fir, red maple, mountain maple, *C. arctata, C. communis, C. intumescens, C. trisperma,* several clubmoss (*Lycopodium*) species, cinnamon fern (*Osmunda cinnamomea*), twisted stalk (*Streptopus roseus*), gold-thread (*Coptis trifolia*), Labrador-tea (*Ledum groenlandicum*).

Similar species. This species has at times been considered a variety or subspecies of *Carex canescens,* but the species are recognized as distinct in modern manuals. Two subspecies (ssp. *sphaerostachya* and ssp. *brunnescens*) are recognized in the state, but they intergrade, and the typical subspecies is the more common. *Carex brunnescens* is similar to *C. canescens* (**silvery sedge [127]**), which generally has *10–20 perigynia per spike* and *spikes longer than wide.*

Inflorescence and perigynium of *Carex brunnescens.*

129. *Carex deweyana*—Dewey's sedge

section *Deweyanae*

After Chester Dewey, a nineteenth-century North American botanist, historian, and educator whose published studies of Carex span 40 years.

The lax inflorescence and narrow, spongy-based perigynia of this species are a common sight in northern Wisconsin, the Baraboo Hills, and counties adjacent to Lake Michigan. Flowers May to June, fruits June to July.

Plants cespitose. **Culms** 0.2–0.9 m tall. **Leaf blades** significantly shorter than culms, the widest 2.5–4 mm wide. **Lowest bract** very narrow, generally 1–5 cm long. **Spikes** gynecandrous, usually somewhat longer than wide, the uppermost often overlapping, *lowest 2 spikes separated by 1–3 cm.* **Perigynia** greenish or brown, *with a prominent whitish, spongy base,* rounded and veinless to few veined on the inner face, veinless on the back, 4–5 mm long, 3–4 times as long as wide; epidermis on the back of the perigynium very thin over the achene, tearing easily in dried specimens; beak tapering, margins finely serrate.

Habitat and state range. Common in northern mesic forests, occasional in northern dry-mesic forests; primarily in northern Wisconsin and the Baraboo Hills, south along Lake Michigan to Milwaukee County. This species is absent from very wet sites where its close relative, *Carex bromoides,* thrives. Common northern mesic forest associates include sugar maple, yellow birch, hemlock, *C. arctata, C. brunnescens, C. communis, C. gracillima, C. leptonervia, C. pedunculata,* wood ferns (*Dryopteris carthusiana* and *D. intermedia*), oak fern (*Gymnocarpium dryopteris*), wild sarsaparilla (*Aralia nudicaulis*), bigleaved aster (*Aster macrophyllus*), blue cohosh (*Caulophyllum thalictroides*), Canada mayflower (*Maianthemum canadense*), wood millet (*Milium effusum*), sweet cicely (*Osmorhiza claytonii*), starflower (*Trientalis borealis*).

Similar species. The inflorescence of Dewey's sedge may superficially resemble the nodding inflorescence of *C. tenera* var. *tenera* (marsh straw sedge [157a]), but *C. tenera* var. *tenera* has *winged perigynia that are not spongy at the base.* Dewey's sedge is closely related to *C. bromoides* (bromelike sedge [130]), which grows in lowland forests, on lake edges, and in cool, springy wetlands in roughly the same geographic range as Dewey's sedge. *Carex bromoides* has *lanceolate perigynia that are usually > 4 times as long as wide.* This gives the spikes a wispy look, easily distinguished from the relatively plump spikes of Dewey's sedge. Moreover, the *perigynium backs in C. bromoides are strongly veined at least to the base of the beak. Carex bromoides* tends to form large clumps that look unkempt by mid- to late summer.

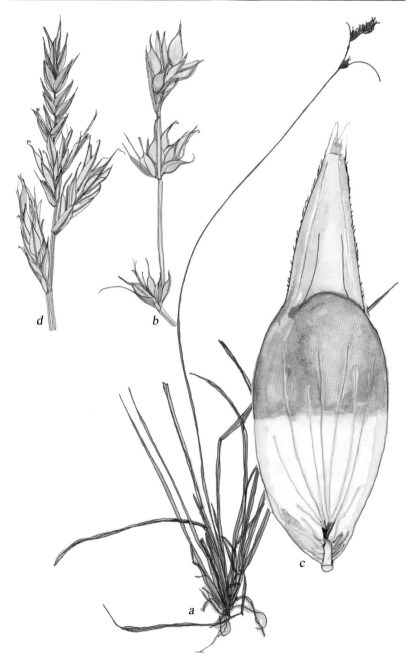

(a) Plant, (b) inflorescence, and (c) perigynium of *Carex deweyana*. (d) Inflorescence of *C. bromoides*. The slender perigynia and grasslike inflorescence of *C. bromoides* are easy to pick out in the field.

133. CAREX INTERIOR—INLAND SEDGE

section *Stellulatae*

Latin: inland.

The starlike spikes of *Carex interior* and its kin recall the upland species *C. rosea* and *C. radiata*, but the clavate base of the terminal spike distinguishes species of this section from *C. rosea* and relatives. Flowers May, fruits June, perigynia falling in late June or July.

Plants cespitose. **Culms** 0.1–1 m. **Leaf blades** erect, grasslike, especially prior to flowering, narrow, the widest blades 1–2.5 mm wide. **Spikes** 2–5 (typically 3), gynecandrous, the terminal spike usually with a clavate, staminate base that is as long as the pistillate portion. **Perigynia** convex and distinctly veined on the back, veinless on the inner face, widest at the base, *apex often rounded and abruptly narrowed to the beak; beak < half as long as the perigynium body.*

Habitat and state range. Common in calcareous wetlands, including sedge meadows, groundwater seeps, wet prairies, fens, and white cedar swamps, occasional in northern sphagnum bogs; concentrated in the eastern half of the state, with scattered populations throughout the western half. Fen and wet prairie associates include *Carex buxbaumii*, *C. conoidea*, *C. pellita*, Kalm's Saint-John's-wort (*Hypericum kalmianum*), swamp saxifrage (*Saxifraga pensylvanica*). White cedar swamp associates include *C. aurea*, *C. disperma*, *C. flava*, *C. leptalea*, *C. tenuiflora*, *C. trisperma*, gold-thread (*Coptis trifolia*), marsh fern (*Thelypteris palustris*).

Similar species. The species most nearly resembling *Carex interior* is *C. echinata* (prickly sedge [134]), which is mostly found in northern acidic wetlands, often in sandy or peaty soils. *Carex echinata* generally has *4 or more spikes* and *perigynia tapering to a beak that is more than half as long as the perigynium body* and generally not as abruptly separated from the perigynium body. *Carex echinata* also often has leaf blades more than 2 mm wide, but leaf blade width on the two species overlaps. *Carex sterilis* (fen star sedge [132]) is a *dioecious* species of fens and calcareous wet prairies. Occasional, mostly staminate plants bear a few perigynia or even whole pistillate spikes, and rare, mostly pistillate plants bear a few staminate flowers.

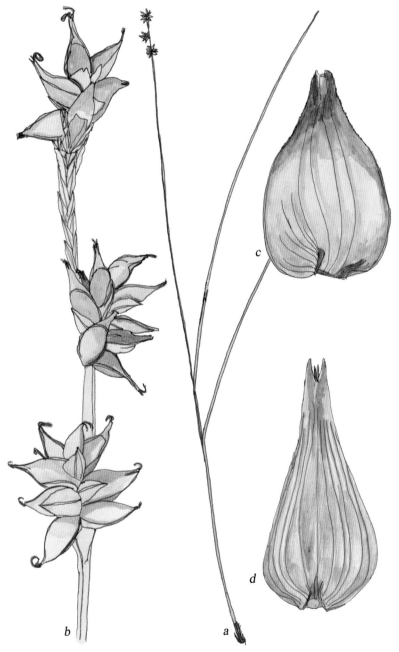

(a) Plant, (b) inflorescence, and (c) perigynium of *Carex interior*. (d) Perigynium of *C. echinata*. The shorter, more abruptly beaked perigynia of *C. interior* set the species apart from *C. echinata*.

136. CAREX MUSKINGUMENSIS—MUSKINGUM SEDGE
section *Ovales*
After Muskingum County, Ohio.

Spindle-shaped spikes and leafy, canelike vegetative culms make Muskingum sedge one of Wisconsin's most charismatic floodplain species. Horticultural varieties of this species are commercially available. Flowers June, fruits July, spikes generally remaining intact well into autumn.

Plants cespitose; *rhizomes stout, blackish, short-creeping.* **Fertile culms** 0.2–1 m tall. **Vegetative culms** more numerous than the fertile culms, *the conspicuously 3-ranked spreading leaves distributed evenly along the upper half to two-thirds of the shoot,* suggesting a *Dulichium* with unnaturally long leaves. **Leaf sheaths** *green-veined on the inner band nearly to the brown-tinged summit,* rounded on the margins. **Leaf blades** *auricle-like at the juncture between the blade and sheath,* 3–6 mm wide. **Inflorescence** stiff or arching, 4–8 cm long. **Spikes 5–12** gynecandrous, overlapping, *narrowly spindle-shaped, tapered to the base and usually to the apex.* **Perigynia** scalelike, 6–9 mm long, strongly veined on both faces; wings narrowed below the midpoint of the body.

Habitat and state range. Floodplain forests, chiefly along the Mississippi, Wisconsin, and Wolf rivers and their major tributaries. Common associates include silver maple, swamp white oak, river birch, American elm, buttonbush, *Carex grayi, C. lupulina, C. tribuloides, C. typhina,* and other floodplain forest species (see additional associates under *C. tribuloides* [138]).

Similar species. The vegetative shoots of *Carex muskingumensis* are unlike anything else in the genus. The putative close relatives *C. tribuloides, C. cristatella,* and *C. projecta* have *sheaths loose and expanded at the apex, the margins acute and becoming winged upon pressing.* Moreover, while the vegetative culms of *C. muskingumensis* appear never to produce sprouts from the nodes, *C. tribuloides* and *C. projecta* reproduce vegetatively from the nodes of fallen vegetative culms in most Wisconsin populations. The turbinate spikes and long, scalelike perigynia of *C. muskingumensis* are unlike any other species in the genus.

Vegetative culm and inflorescence of *Carex muskingumensis*. The spindle-shaped spikes remain intact well into the fall, so that the plant is readily identified vegetatively or in fruit as late in the season as any other *Carex* in our flora.

138. *CAREX TRIBULOIDES*—AWL-FRUITED OVAL SEDGE
section *Ovales*

After the genus *Tribulus* (Zygophyllaceae), the fruits of which have sharp protrusions that point outward in a manner similar to the four spikes of a caltrop.

The leafy vegetative culms with baggy sheaths are characteristic of the three closely related species discussed here. Flowers May to June, fruits July, perigynia frequently persisting into fall.

Plants cespitose; rhizomes short. **Vegetative culms** 0.2–1 m tall, more numerous than the fertile culms, producing numerous spreading leaves on the upper half, *reproducing in part by means of plantlets produced at the nodes of the previous year's vegetative culms, resembling stolons.* **Leaf sheaths** *loose, expanded near the summit; inner band green-veined nearly to the summit, firm.* **Leaf blades** 3–7 (occasionally 2) mm wide. **Inflorescence** stiff, rarely arching. **Spikes** gynecandrous, overlapping, *tapered to the base, apex typically rounded in mature material, tapered in immature spikes and some mature spikes.* **Perigynia** ≥ 30 per spike, lanceolate, thin and scalelike, 3–5.5 mm long, 1–1.5 mm wide; inner face distinctly veined and translucent, revealing the dark achene; *wings narrowed below the midpoint.*

Habitat and state range. Most common in floodplain forests of larger rivers, ranging to prairie sloughs, wet fields, sedge meadows, and sandbars of the Wisconsin River; more or less restricted to alluvial soils in the southern two-thirds of the state. Common associates include floodplain species such as *Carex lupulina, C. muskingumensis, C. tuckermanii, C. typhina,* wild yam (*Dioscorea villosa*), rice cut-grass (*Leersia oryzoides*), cardinal flower (*Lobelia cardinalis*), moneywort (*Lysimachia nummularia*), obedience plant (*Physostegia virgiana*), clearweed (*Pilea pumila*), cut-leaf coneflower (*Rudbeckia laciniata*).

Similar species. *Carex tribuloides* is most frequently confused with *C. projecta* (necklace sedge [139]), which is common in mesic to wet woodlands and wooded edges in the northern two-thirds of the state. *Carex projecta* usually has a *loose, arching inflorescence; fewer than 30 perigynia per spike;* and *perigynium tips at the spike apex arching outward (spreading).* *Carex cristatella* (crested oval sedge [137]), a floodplain species primarily of the eastern third and southern half of the state but scattered throughout most of the state, excluding the northern highlands, differs from both of these in having *nearly spherical spikes* and *strongly reflexed perigynium beaks, the pistillate scales concealed by the perigynia at maturity.*

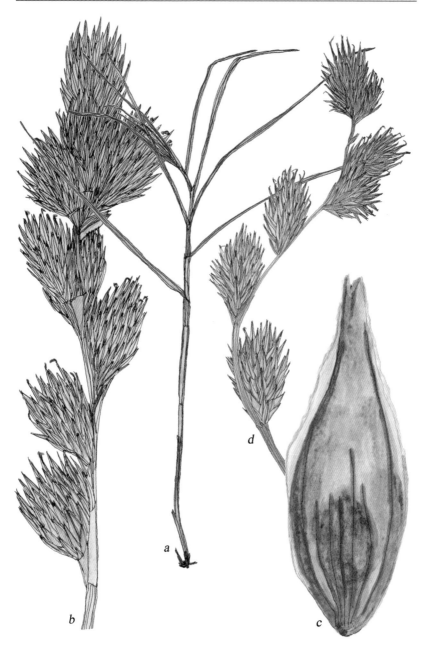

(a) Vegetative culm, (b) inflorescence, and (c) perigynium of *Carex tribuloides*. (d) Inflorescence of *C. projecta*. The more open and flexuous inflorescence of *C. projecta* usually distinguishes it from both *C. tribuloides* and *C. cristatella*, but some individuals are not easily identified.

147. *Carex bicknellii*—Bicknell's sedge

section *Ovales*

After Eugene Pintard Bicknell (1859–1925), a New York businessman and amateur botanist.

The brown perigynium margins of this species are utterly distinctive. Flowers May to June, fruits June to July, perigynia mostly dispersing in August or September.

Plants cespitose. **Culms** 0.2–1 m tall. **Leaf blades** 2–5 mm wide. **Inflorescence** erect to arching, occasionally congested, more often open. **Spikes** gynecandrous, ≥ 4 per inflorescence, usually tapered to the base; apex prickly with perigynium beaks. **Pistillate scales** coppery or brownish. **Perigynia** *coppery brown on the upper margins at maturity, translucent, the dark achene visible through the perigynium wall*, 4.5–7 mm long, 2.5–4 mm wide, conspicuously veined on both faces. **Anthers** ≥ 2.5 mm long.

Habitat and state range. Most common in mesic to wet-mesic prairies in the southern half of the state, with a few populations in northwestern Wisconsin. Typical associates include *Carex pellita*, big bluestem (*Andropogon gerardii*), rattlesnake master (*Eryngium yuccifolium*), grass-leaved goldenrod (*Euthamia graminifolia*), prairie dropseed (*Sporobolus heterolepis*), Culver's root (*Veronicastrum virginicum*).

Similar species. Bicknell's sedge is most similar to **Carex merritt-fernaldii** (**Fernald's sedge [148]**), a species of dry sandy soils in northern Wisconsin. Perigynia of Fernald's sedge are ≥ 2.5 mm wide and translucent, like those of *C. bicknellii*, but *shorter* (3.5–5 mm long), *faintly veined or unveined on the inner face*, and *yellow on the margins* at maturity, anthers typically ≤ 2.5 mm long. **Carex brevior** (**fescue sedge [151]**), of dry to mesic southern Wisconsin prairies, suggests a diminutive *C. bicknellii* with *smaller perigynia* (≤ 5 mm long) *veinless on the inner face*, the relatively opaque perigynium bodies nearly circular in outline (orbiculate). Pistillate scales in this species generally extend to the middle or tip of the perigynium beak. **Carex molesta** (**troublesome sedge [149]**), common in southern Wisconsin low to wet-mesic prairies and wetland edges, resembles *C. brevior*, but with *congested inflorescences of 3–4 nearly spherical spikes*, packed together like tiny burs at the tip of the culm. *Perigynia in C. molesta are more nearly elliptical than orbiculate* and *veined on the inner face*, and the *pistillate scales generally extend only to the base of the beak.*

(a) Plant, (b) inflorescence, and (c) perigynium of *Carex bicknellii*. The achene is visible through the translucent epidermis of the perigynium in this illustration, as it is in the field and herbarium. (d) Perigynium of *C. brevior*. The orbiculate perigynium body is distinctive.

152. *CAREX CRAWFORDII*—CRAWFORD'S SEDGE

section *Ovales*

After Ethan Allen Crawford, the New Hampshire mountaineer who helped carve the first footpath to the summit of Mount Washington in the early nineteenth century.

This species is reminiscent of *Carex bebbii,* but with lanceolate perigynia. Flowers June, fruits June to July, perigynia usually dispersing in August.

Plants cespitose. **Culms** wiry, 0.1–1 m tall. **Leaf blades** 2–4 mm wide. **Inflorescence** compact, usually with 1–2 bristlelike bracts at the base, 2–3 cm long, 1–1.5 cm thick. **Spikes** gynecandrous, usually tapered to an *acute base; apex generally bristly with elongate perigynium tips.* **Perigynia** *lanceolate, 3.5–4 mm long, approximately 1 mm wide,* scalelike except over the achene, frequently veined on the inner face; beak long-tapering.

Habitat and state range. Most characteristic of sandy open wetlands such as lakeshores and sedge meadows with such associates as *Carex comosa, C. crinita, C. debilis* var. *rudgei, C. hystericina, C. intumescens, C. scoparia;* frequent in disturbed roadsides and waste areas; occasional in sandy uplands. Common and widespread in northern Wisconsin.

Similar species. Crawford's sedge may be mistaken for **Carex scoparia** (broom sedge [153]), but the exceedingly narrow perigynia of the former are generally distinguishable from broom sedge's *wider perigynia* (1.2–2 mm wide). Crawford's sedge may also be confused with **C. bebbii** (Bebb's sedge [154]), which rarely has lanceolate perigynia or beaks as elongate as those of Crawford's sedge.

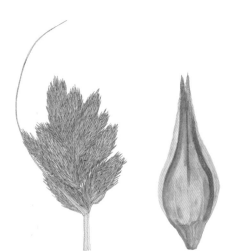

Inflorescence and perigynium of *Carex crawfordii.*

153. *CAREX SCOPARIA*—BROOM SEDGE

section *Ovales*

Latin: broom.

This is one of Wisconsin's most common *Ovales* species and one of its most variable; Fernald recognized five forms in *Carex scoparia* var. *scoparia*. The 1.2–2 mm wide lanceolate to ovate perigynia, spikes that are often tapered to both ends, and tight leaf sheaths distinguish this species from similar species in the section. Flowers May to June, fruits June to July.

Plants densely cespitose, often forming large clumps, sometimes with conspicuous, leafy vegetative culms. **Culms** 0.2–1 m tall. **Leaf sheaths** tight. **Leaf blades** 1.5–3.5 mm wide. **Inflorescence** compact to slender and elongate, if long, then typically arching, 1.5–6 cm long, *variable within and between populations*. **Spikes** gynecandrous, overlapping, tapered to an acute base and an acute or rounded apex. **Perigynia** lanceolate to ovate, generally scalelike except over the achene, 4–7 mm long, 1.2–2 mm wide, conspicuously veined on both faces.

Habitat and state range. Most common in open, wet, sandy soil; occasionally in microsites that are more or less bare of other vegetation. The species ranges from shallow water (base of plant submerged) to, rarely, dry sandy uplands and from sun to partial shade. Common in wetlands throughout most of the state, uncommon in east-central Wisconsin, Door County, and much of the Driftless Area (though present there in floodplains). Typical habitats include marshes, sedge meadows, lakeshores, ditches, wet prairies, and occasional sphagnum bogs, with associates such as *Carex crawfordii, C. echinata, C. tenera, C. utriculata.*

Similar species. Broom sedge is sometimes confused with *Carex tribuloides* (awl-fruited oval sedge [138]), but the loose sheaths, wider leaves, and prominent vegetative culms of the latter distinguish it. Moreover, the inflorescence of *C. tribuloides* is almost always stiff and elongate, while that of *C. scoparia* is compact to elongate and somewhat drooping. See also *C. crawfordii* (Crawford's sedge [152]).

Inflorescence and perigynium of *Carex scoparia.*

154. *CAREX BEBBII*—BEBB'S SEDGE

section *Ovales*

After the botanist and willow specialist Michael Schuck Bebb (1833–1895).

This common wetland species with a congested inflorescence and small perigynia that are largely veinless on the inner face is usually easily recognized, though it is sufficiently variable to present difficulties in some cases. Flowers June, fruits June to July.

Plants densely cespitose, vegetative culms sometimes evident but usually not conspicuous, leaves clustered at the apex rather than distributed along the length of the shoot. **Culms** 0.3–1 m tall. **Leaf sheaths** tight, not expanded at the summit. **Leaf blades** roughly equaling the tips of the inflorescence, ≤ 4 mm wide. **Inflorescence** *congested, approximately 1–2 times as long as wide;* lowest bract frequently conspicuous, bristlelike. **Spikes** gynecandrous, brown at maturity; base frequently rounded to obtuse; apex rounded. **Perigynia** elliptical, 2.5–4 mm long, 1.2–2 mm wide, *inner face veined faintly or only near the base.*

Habitat and state range. Calcareous wetlands, including low prairies, fens, marshes, ditches, stream edges, lakeshores, and wet old fields, ranging occasionally to white cedar swamps and more upland areas primarily in the eastern half of the state, some populations in the western half. Associates include *Carex conoidea, C. cristatella, C. crawei, C. garberi, C. interior, C. tenera, C. vulpinoidea.*

Similar species. Bebb's sedge may resemble *Carex normalis* (**greater straw sedge [155]**), *C. tincta* (**tinged oval sedge [156]**), or the occasional *C. tenera* var. *tenera* (**marsh straw sedge [157a]**) with a congested inflorescence. The lack of obvious veins on the inner perigynium face of the perigynium usually distinguishes *C. bebbii* from these species. Bebb's sedge is also sometimes confused with *C. cristatella* (**crested oval sedge [137]**), but the latter's *prominent vegetative culms with baggy sheaths and broad leaves*—at least some in the population usually wider than 6 mm—generally distinguish it.

Inflorescence and perigynium of *Carex bebbii.*

155. *CAREX NORMALIS*—GREATER STRAW SEDGE
section *Ovales*
Latin: right-angled.

Stiff inflorescence, broad leaf blades, and whitened areas between the veins on the backs of the leaf sheaths distinguish this species from Wisconsin's other upland members of section *Ovales*. A form with slender, elongate, arching inflorescences crops up sporadically throughout the range of the species (f. *perlonga*). Flowers June, fruits June to July.

   Plants cespitose, vegetative culms sometimes conspicuous, leaves clustered at the apex. **Leaf sheaths** *conspicuously whitened on the backs between the veins;* inner band hyaline or green-veined, *prolonged at the summit slightly beyond the juncture with the base of the leaf blade.* **Leaf blades** 2–6 mm wide. **Inflorescence** typically compact, straight, very occasionally elongate and arching, with spikes distant from one another (*C. normalis* f. *perlonga*), 1.5–5 cm long. **Spikes** overlapping or the lowest separate. **Perigynia** spreading, greenish, 2.5–4 mm long, 1.5–2.3 mm wide, distinctly veined on both faces.

   Habitat and state range. Mostly in mesic forests, ranging to dry woodlands and savannas, forest edges, and occasional mesic prairies and wetland margins; tolerant of disturbance. Throughout the state, more common in the southern half. Common associates include sugar maple, red oak, white oak, *Carex cephalophora*, *C. pensylvanica*, *C. tenera*, hog-peanut (*Amphicarpaea bracteata*), rattlesnake fern (*Botrychium virginianum*), woodland tick-trefoil (*Desmodium glutinosum*), sweet cicely (*Osmorhiza* spp.).

   Similar species. *Carex normalis* is most readily confused with **C. *tenera* var. *echinodes*** (quill sedge [157b]), which differs in having its *inflorescence typically nodding from beneath the lowest spike, perigynium beaks spreading,* and *leaves generally narrower.* However, wide-leaved forms of *C. tenera* var. *echinodes* can be difficult to distinguish. The less closely related **C. *cephaloidea*** (clustered bracted sedge [114]) is superficially similar but has *spikes androgynous.*

Inflorescence of *Carex normalis.*

157A. *CAREX TENERA* VAR. *TENERA*—MARSH STRAW SEDGE

section *Ovales*

Latin: tender or soft, perhaps referring to the foliage.

This species' slender, gracefully drooping inflorescence and small spikes are highly distinctive. Flowers May to June, fruits June to July, perigynia mostly falling in August.

Plants cespitose. **Culms** 0.2–1 m tall, leaning or arching, narrow. **Leaf sheaths** tight, usually green on the backs, papillose (30× magnification). **Leaf blades** 1.5–2.5 mm wide. **Inflorescence** open (very occasionally congested), *nodding from beneath the lowest spike*. **Spikes** gynecandrous, mostly not overlapping, base tapered or clavate. **Pistillate scales** usually extending to middle of perigynium beak. **Perigynia** roughly elliptical, 3–4 mm long, veined on the inner face; beak tips exceeding tips of the pistillate scales by ≤ 0.8 mm.

Habitat and state range. Most common in mesic to wet prairies, sedge meadow borders, and dry to mesic oak–hickory woodlands; very occasional in mesic forests (which are more commonly inhabited by *Carex tenera* var. *echinodes*) and dry prairies throughout the state. Common wet prairie associates include *C. bebbii, C. bicknellii, C. buxbaumii, C. conoidea, C. pellita, C. scoparia,* bluejoint grass (*Calamagrostis canadensis*), blue bottle gentian (*Gentiana andrewsii*), prairie dock (*Silphium terebinthinaceum*), prairie dropseed (*Sporobolus heterolepis*).

Similar species. *Carex festucacea* (fescue sedge [150]), *C. projecta* (necklace sedge [139]), and *C. straminea* (straw sedge [143]) have nodding inflorescences that superficially resemble those of *C. tenera*. *Carex projecta* is distinguished by its *loose leaf sheaths, prominent vegetative culms,* and *leaves > 3 mm broad*; the other two species are distinguished by having *perigynium bodies roughly orbiculate*. The closely related **C. *tenera* var. *echinodes*** (straw sedge [157b]), a wet to mesic forest taxon distributed mostly in the southern half of the state, produces nodding inflorescences similar to those of *C. tenera* var. *tenera*, but the *perigynium tips arch outward and often exceed the pistillate scale tips by ≥ 1 mm*. With 30× magnification, this variety can also be distinguished from *C. tenera* var. *tenera* by its *smooth (not papillose) leaf sheaths*.

Plant, inflorescence, and perigynium of *Carex tenera* var. *tenera*.

Appendix A

Principal *Carex* Habitats of Wisconsin
and Their Typical Species

THEODORE S. COCHRANE

The following outline presents the principal sedge habitats in Wisconsin and lists their typical and rare species. Its purposes are to associate our sedges with the places in which they normally grow and to help those studying sedges identify their specimens more easily by restricting the number of species under consideration. All taxa are taken into account, including our few adventive and introduced species, so that each appears at least once. State-listed endangered, threatened, and special concern species appear under the subheading "Rare species" (see Wisconsin Department of Natural Resources 2004). The categories in which our carices have been placed avoid reference to any system of precisely defined ecological communities; instead, they are the characteristic habitats in which our sedges are to be found, summarized in broad terms that will be understandable to everyone. The habitats are grouped under the terms "Southern Wisconsin" and "Northern Wisconsin," areas situated south or north of the climatic and floristic Tension Zone (see Curtis 1959) and thus approximately defined as lying along either side of a line running from St. Croix Falls in the northwest to Manitowoc in the east.

SOUTHERN WISCONSIN

Dry open ground: Habitat 1

Dry prairies (grading into mesic prairies), sand barrens (open or with black oak or jack pine), oak openings, and "cedar glades"; exposed cliffs, bluffs, rocks, and banks

Typical species

C. *albicans* var. *emmonsii* (acid sand)
C. *bicknellii*
C. *brevior*
C. *eburnea* (usually under red or white cedar)
C. *gravida* (more often in old fields and on roadsides)
C. *inops*

C. *meadii*
C. *muehlenbergii* var. *enervis* (sand)
C. *muehlenbergii* var. *muehlenbergii* (sand)
C. *pensylvanica* (sometimes the dominant ground cover)
C. *siccata* (sand)
C. *tonsa* var. *rugosperma* (usually in sand)

C. *tonsa* var. *tonsa* (sand)
C. *umbellata*

Occasional southward

C. *foenea* (less often in cool woods)

Rare species

C. *albicans* var. *albicans* (rocks; Sauk
County)
C. *richardsonii* (also on beach dunes)
C. *torreyi* (mesic prairies; less often in
moist oak woods)

Dry open ground: Habitat 2

Meadows, fields, roadsides, and lawns; hillsides, banks, and borders of woods; ruderal sites

Typical species

C. *aggregata* (crack in railroad tie;
adventive from farther south)
C. *annectens*
C. *bicknellii*
C. *blanda* (generally in woods)
C. *bushii* (introduced from farther south)
C. *cephalophora* (more often in woods)
C. *gravida* (seldom in prairies)
C. *leavenworthii* (introduced from farther
south)

C. *muehlenbergii* var. *enervis* (sand)
C. *muehlenbergii* var. *muehlenbergii*
(sand)
C. *normalis* (usually in woods and thickets)
C. *pellita* (characteristically in wet ground)
C. *praegracilis* (introduced from farther
west)
C. *scoparia* (normally in moist ground)
C. *siccata* (sand)
C. *spicata* (introduced from Europe)

Wet open ground: Habitat 3

Prairies, low meadows, damp fields, and fens; sandy, peaty, or marly shores and stream-sides; beach flats and interdunal swales

Typical species

C. *alopecoidea*
C. *annectens*
C. *aquatilis* var. *aquatilis*
C. *aquatilis* var. *substricta*
C. *atherodes* (usually in shallow water and
on shores)
C. *aurea* (damp sand, marl, or clay, mostly
along the Great Lakes)
C. *bebbii*
C. *bicknellii*
C. *bromoides*
C. *buxbaumii*
C. *conoidea*
C. *cristatella*
C. *cryptolepis*
C. *debilis*
C. *diandra* (quaking or floating sedge mats)
C. *festucacea*
C. *flava*
C. *granularis*
C. *gynandra*
C. *haydenii*

C. *hystericina* (slightly disturbed peat)
C. *interior*
C. *lacustris*
C. *laeviconica* (along the Mississippi and
lower Wisconsin rivers)
C. *lasiocarpa* (usually on sedge mats and
in open bogs)
C. *leptalea* (more often in tamarack bogs)
C. *lurida* (wet acid site near streams)
C. *molesta*
C. *normalis*
C. *pellita*
C. *prairea*
C. *sartwellii* (characteristically in sedge
meadows)
C. *scoparia*
C. *sterilis*
C. *stipata*
C. *stricta* (dominant ground cover in sedge
meadows)
C. *tenera* var. *tenera*
C. *tetanica*

C. trichocarpa (stream valleys)
C. utriculata (usually shallow water and
shores)
C. viridula (sandy lakeshores*)*
C. vulpinoidea

Rare species

C. crawei
C. cumulata (moist sand in burned-over or
scraped-off bogs)

C. longii (sandy sphagnous meadow; La
Crosse County)
C. richardsonii (characteristic of dry
prairies)
C. straminea (sandy sedge meadows and
swamp borders)
C. suberecta (sedge meadows)
C. sychnocephala (drying sandy or mucky
shores)

Wet open ground: Habitat 4

Marshes, wet shores, and shallow water along lakes, ponds, and streams; quaking or floating sedge mats; edges of and openings in swamps and bogs; ditches

Typical species

C. alopecoidea (almost always on
floodplains)
C. annectens
C. aquatilis var. *aquatilis*
C. aquatilis var. *substricta*
C. atherodes (often on margins of cattail
marshes)
C. buxbaumii
C. canescens (characteristically in bogs)
C. comosa
C. diandra (especially springy places and
floating mats)
C. emoryi (characteristically on wooded
river floodplains)
C. hirta (marshy area around pond;
introduced from Eurasia)
C. hystericina
C. interior
C. lacustris
C. laeviconica (along the Mississippi and
lower Wisconsin rivers)

C. lasiocarpa
C. lupulina
C. lurida (wet acid sites near streams)
C. nebrascensis (railroad ditch; adventive
from farther west)
C. oligosperma
C. pellita
C. prairea
C. pseudocyperus (often in shallow water)
C. retrorsa
C. stipata
C. stricta
C. trichocarpa
C. utriculata
C. vesicaria
C. vulpinoidea

Rare species

C. folliculata (peaty ditches; usually in
swamps and wet thickets)

Wet open ground: Habitat 5

Sphagnum bogs (relics in the south) and conifer swamps (usually in extinct lake beds in
central Wisconsin)

Typical species

C. brunnescens
C. chordorrhiza
C. cryptolepis
C. diandra
C. echinata

C. lacustris
C. lasiocarpa
C. leptalea
C. limosa
C. magellanica

C. oligosperma
C. pseudocyperus
C. trisperma
C. utriculata

Rare species

C. livida
C. tenuiflora (mostly north of the Tension
   Zone)

Dry shaded ground: Habitat 6

Upland deciduous forests, including maple-basswood stands (with or without beech), moist
oak to dry oak-hickory woods, open woodlands, and disturbed woodlots

Typical species

C. albursina (especially on mounds, slopes,
   and paths)
C. blanda
C. cephaloidea
C. cephalophora
C. communis (especially in clearings and
   disturbed spots)
C. deweyana
C. digitalis
C. gracillima
C. granularis
C. grisea
C. hirtifolia
C. hitchcockiana (maple-basswood forests)
C. inops
C. jamesii (maple forests)
C. laxiculmis var. *copulata*
C. laxiculmis var. *laxiculmis*
C. laxiflora
C. normalis
C. oligocarpa (maple-basswood forests)
C. peckii
C. pedunculata (more common northward)

C. pensylvanica
C. plantaginea
C. radiata
C. rosea
C. sparganioides
C. sprengelii
C. woodii (sugar maple forests)

Occasional southward

C. leptonervia

Rare species

C. albicans var. *albicans* (quartzite bluffs;
   Sauk County)
C. backii (cool sandy slopes or cliffs)
C. careyana
C. gracilescens
C. media (algific talus slope; Grant County)
C. prasina (woodland springs)
C. swanii
C. torreyi (moist oak woods; more often
   in prairies)

Wet shaded ground: Habitat 7

Lowland deciduous forests, swampy, boggy, or springy woods, and wet thickets; wooded
streamsides, floodplains, and swamp borders; mucky hollows and depressions in woods

Typical species

C. albicans var. *emmonsii* (low sphagnous
   woods)
C. alopecoidea
C. blanda
C. bromoides
C. brunnescens (local southward)
C. crinita (especially borders and clearings)
C. cristatella

C. davisii (along larger rivers)
C. debilis
C. emoryi (along streams)
C. gracillima
C. granularis
C. grayi (along larger rivers)
C. grisea
C. gynandra

C. hirtifolia
C. hitchcockiana
C. intumescens
C. jamesii
C. lupulina
C. lurida (low sphagnous woods bordering lakes and streams)
C. molesta
C. muskingumensis
C. projecta
C. retrorsa
C. stipata
C. tenera var. *echinodes*
C. tribuloides (often along rivers)
C. tuckermanii
C. typhina
C. utriculata
C. vesicaria

C. vulpinoidea
C. woodii

Occasional southward

C. canescens var. *canescens*
C. canescens var. *disjuncta*

Rare species

C. crus-corvi
C. folliculata (white pine–red maple swamps and alder thickets)
C. laevivaginata
C. lupuliformis (alluvial woods)
C. schweinitzii (cold springy areas; Iowa County)
C. straminea (swamp borders and sandy sedge meadows)

Northern Wisconsin

Dry open ground: Habitat 8

Sandy, gravelly, or rocky ground, including barrens and dunes, ridges and outcrops, and disturbed sites (old fields, lawns, roadsides, and railroads)

Typical species

C. adusta
C. foenea
C. houghtoniana
C. inops
C. merritt-fernaldii
C. normalis (usually in moist ground)
C. ovalis (Eurasian; introduced in the Apostle Islands)
C. pensylvanica
C. praegracilis (introduced from farther west)
C. scoparia (characteristically in moist ground)

C. siccata
C. tonsa var. *rugosperma*

Occasional northward

C. bicknellii
C. brevior
C. muehlenbergii var. *muehlenbergii*
C. umbellata

Rare species

C. richardsonii
C. tincta (Ashland County)

Wet open ground: Habitat 9

Damp to wet sites of all kinds except sphagnum bogs: shallow water, marshes, edges of ponds, lakes, rivers, and streams; beach pools and swales; ditches; meadows, fields, clearings, and other openings

Typical species

C. aquatilis (mostly var. *substricta*)
C. bebbii
C. bromoides

C. buxbaumii
C. canescens ssp. *canescens* (usually in bogs and swamps)

C. comosa
C. crinita
C. cristatella
C. cryptolepis
C. debilis
C. diandra
C. echinata
C. flava
C. granularis
C. gynandra
C. haydenii
C. houghtoniana (usually in dry
    ground)
C. hystericina
C. interior
C. lacustris
C. lasiocarpa
C. leptalea (often in bogs and swamps)
C. oligosperma (usually in sphagnum
    bogs)
C. pellita
C. prairea
C. projecta (usually in swamps and
    thickets)
C. pseudocyperus

C. retrorsa
C. rostrata
C. sartwellii
C. scoparia
C. sterilis
C. stipata
C. stricta
C. tenera var. tenera
C. tuckermanii (characteristically in
    swamps and thickets)
C. utriculata
C. vesicaria
C. vulpinoidea

Occasional northward

C. annectens
C. atherodes
C. laeviconica
C. tetanica
C. tribuloides
C. trichocarpa

Rare species

C. nigra
C. pallescens

Wet open ground: Habitat 10

Damp sandy, gravelly, or marly shores (occasionally in marshes, meadows, ditches, and other moist open sites)

Typical species

any of the Habitat 9 species, especially
    C. hystericina, C. scoparia, C. stiptata,
    C. vulpinoidea, plus:
C. adusta (usually in dry sandy or rocky
    barrens)
C. aurea (chiefly near the Great Lakes)
C. crawfordii
C. eburnea (usually under conifers)

C. echinata
C. viridula (sandy lakeshores)

Rare species

C. crawei
C. garberi (Lake Michigan shores)
C. lenticularis
C. sychnocephala (drying mud and mucky
    sand)

Wet open ground: Habitat 11

Open sphagnum bogs (very closely related to conifer bogs and swamps), including sparsely wooded tamarack–black spruce–sphagnum bogs, treeless glades in black spruce and/or tamarack forests, floating sedge and sphagnum–shrub mats, and peaty interdunal swales

Typical species

C. aquatilis var. aquatilis
C. aquatilis var. substricta

C. brunnescens (usually in conifer swamps)
C. canescens ssp. canescens

C. *canescens* ssp. *disjuncta*
C. *chordorrhiza*
C. *comosa*
C. *cryptolepis* (usually in marshes, shores, ditches)
C. *diandra*
C. *echinata*
C. *flava*
C. *interior*
C. *lasiocarpa*
C. *leptalea*
C. *limosa*
C. *magellanica*
C. *oligosperma*

C. *pauciflora*
C. *pseudocyperus*
C. *rostrata*
C. *sterilis*
C. *utriculata*

Rare species

C. *exilis* (beach-pool bog mats; Ashland County)
C. *livida* (boreal fens and calcareous floating mats)
C. *michauxiana*
C. *tenuiflora*
C. *vaginata*

Dry shaded ground: Habitat 12

Dry jack pine (with Hill's oak and poplar) and red pine (with red maple) forests; aspen woodlands; wooded dunes, sandy banks, rocky hillsides, and bluff tops

Typical species

C. *arctata* (characteristic of rich woods)
C. *communis*
C. *deflexa* (often on knolls or along trails)
C. *deweyana*
C. *inops*
C. *lucorum*
C. *merritt-fernaldii* (usually in open ground)
C. *ormostachya*
C. *peckii* (usually in maple-hemlock woods)
C. *pedunculata* (characteristic of rich woods)

C. *pensylvanica*
C. *rosea*
C. *siccata*
C. *tonsa* var. *rugosperma*
C. *tonsa* var. *tonsa*

Occasional northward

C. *umbellata* (sandy or gravelly ground; characteristic of dry to mesic prairies)

Rare species

C. *backii*

Dry shaded ground: Habitat 13

Rich deciduous or mixed forests, especially beech-maple, maple-basswood, and hemlock-hardwood stands; upland oak woods and other rich woods and thickets

Typical species

C. *albursina*
C. *arctata*
C. *blanda*
C. *castanea*
C. *cephaloidea*
C. *communis* (especially in clearings and along trails)
C. *deflexa* (especially on knolls and along trails)

C. *deweyana*
C. *disperma* (usually in bogs and conifer swamps)
C. *foenea* (often in dry open ground and rocky barrens)
C. *gracillima*
C. *grisea*
C. *hirtifolia*
C. *inops*

C. laxiflora
C. leptonervia
C. normalis
C. ormostachya
C. peckii
C. pedunculata
C. pensylvanica
C. plantaginea
C. radiata
C. rosea
C. sparganioides
C. sprengelii
C. tenera var. *tenera*
C. woodii (sugar maple forests)

Occasional northward

C. cephalophora (usually in dry deciduous forests)
C. laxiculmis var. *laxiculmis*

Rare species

C. assiniboinensis
C. backii (cool slopes, cliffs, or bluffs)
C. formosa
C. gracilescens
C. platyphylla (Door County)
C. prasina (woodland springs)

Dry shaded ground: Habitat 14

Dryish to moist coniferous woods and thickets (especially hemlock, balsam fir, white spruce, red or white pine)

Typical species

C. arctata
C. brunnescens (usually in bogs and swamps)
C. castanea
C. communis (usually in deciduous woods)
C. debilis (usually under hardwoods)
C. deflexa (often on knolls or along trails)
C. deweyana
C. eburnea
C. gracillima

C. inops
C. intumescens
C. pedunculata
C. pensylvanica

Rare species

C. concinna
C. novae-angliae (along alder-lined streams in mixed white pine or balsam fir woods)

Wet shaded ground: Habitat 15

Wet woods and thickets, including swamp forests (chiefly hardwoods rather than conifers, such as black ash–yellow birch–hemlock hardwoods and elm-maple swamps), and wooded riverbanks and floodplains (relatively rare in northern Wisconsin)

Typical species

C. bromoides
C. brunnescens (usually near bogs and in conifer swamps)
C. canescens ssp. *canescens* (usually in bogs and conifer swamps)
C. crinita
C. cristatella
C. emoryi (river and creek floodplains)
C. gracillima
C. granularis (especially along trails and clearings)

C. gynandra
C. intumescens
C. lupulina
C. lurida (wet acid sites near streams)
C. projecta
C. radiata
C. retrorsa
C. scabrata
C. sprengelii
C. stipata
C. trisperma

C. tuckermanii
C. utriculata
C. vesicaria
C. vulpinoidea

Occasional northward

C. alopecoidea
C. grisea
C. hitchcockiana

Rare species

C. assiniboinensis (river floodplain forests)

Wet shaded ground: Habitat 16

Wet spots (borders of ponds, streambanks, seepy areas, depressions) in upland woods

Typical species

C. bromoides
C. crinita
C. gracillima
C. intumescens
C. lupulina
C. projecta
C. retrorsa

C. tuckermanii
C. vulpinoidea

Rare species

C. prasina (woodland seeps, springs, and
   streams)

Wet shaded ground: Habitat 17

Roadsides, trails, clearings, and other openings in or borders of swamps

Typical species

C. crinita (more often in conifer swamps)
C. gracillima
C. granularis
C. gynandra

C. retrorsa
C. tenera var. tenera
C. vulpinoidea

Wet shaded ground: Habitat 18

Conifer swamps, bogs, and thickets, including tamarack–black spruce–sphagnum bogs, older black spruce–tamarack forests, white cedar–balsam fir swamps, mixed conifer stands (with white cedar, balsam fir, jack pine), boggy mixed woods (with hemlock and/or black ash), fir woods, and cedar thickets

Typical species

C. aurea (mostly near the Great Lakes)
C. bebbii
C. brunnescens
C. canescens ssp. canescens
C. canescens ssp. disjuncta
C. castanea
C. comosa
C. cristatella
C. disperma
C. eburnea (usually under red cedar or
   white cedar and/or balsam fir)

C. echinata
C. flava
C. gracillima
C. granularis
C. interior
C. intumescens (more often in deciduous
   woods and thickets)
C. lacustris
C. lasiocarpa (characteristically in open
   bogs)
C. leptalea

C. leptonervia (usually in upland woods)
C. magellanica
C. oligosperma (characteristically in open
 bogs)
C. pauciflora
C. peckii (usually in upland woods)
C. pedunculata
C. pseudocyperus
C. retrorsa
C. sterilis
C. trisperma

Occasional northward

C. prairea

Rare species

C. capillaris
C. crawei (characteristically in beach flats
 and on swales)
C. gynocrates
C. tenuiflora

Wet shaded ground: Habitat 19

Wet sites similar to bogs: borders of bogs, boggy shores and stream margins, swampy thickets and ditches adjacent to sphagnum bogs and conifer swamps; openings (trails, roadsides, clearings, hollows), hummocks, and logs in swampy woods

Typical species

(also some bog species, e.g., *C.
 brunnescens, C. canescens, C. disperma,
 C. gynocrates, C. interior, C. leptalea,
 C. magellanica, C. trisperma*)
C. arcta
C. arctata
C. aurea
C. bebbii
C. castanea
C. comosa
C. crawfordii
C. crinita
C. eburnea
C. echinata
C. gracillima
C. granularis
C. gynandra
C. intumescens

C. lacustris
C. leptonervia
C. lupulina (usually in deciduous swamps
 and thickets)
C. pedunculata
C. retrorsa
C. scabrata (usually in deciduous
 woods)
C. stipata
C. stricta
C. tuckermanii
C. utriculata
C. vesicaria
C. viridula (frequently on shores)
C. vulpinoidea

Rare species

C. capillaris

Appendix B

Atlas of the Wisconsin *Carex* Flora

MEREL R. BLACK AND THEODORE S. COCHRANE

The Wisconsin distribution of each *Carex* species, subspecies, or variety is shown by a map. These maps are a joint product of the Wisconsin Botanical Information System database, created and maintained at the University of Wisconsin–Madison Herbarium (WIS), of the University of Wisconsin–Madison, and handmade maps produced over a period of 30 years by Theodore S. Cochrane. Both the database maps and the Cochrane floristic atlas maps have their roots in the work of Norman C. Fassett, Hugh H. Iltis, and their colleagues, who authored the Preliminary Reports on the Flora of Wisconsin series and books on Wisconsin ferns, grasses, and legumes. These works are based only on specimens examined, and all of them include a map for each species, showing its known distribution by dots, circles, or triangles and implying its relative frequency by the number of symbols. Throughout the preparation of these *Carex* maps, the specimen-based and exact-locality principles have been followed. The names on the maps reflect the same recent taxonomic revisions and nomenclatural decisions that appear in *Flora of North America*, vol. 23, and the Wisconsin Vascular Plants website (www .botany.wisc.edu/wisflora), except in three cases in which it proved difficult to recognize taxa at the level of subspecies. The arrangement of taxa follows the numbering system used throughout this book.

In 1974 Cochrane launched a long-term effort to gather sedge data from as many Wisconsin herbaria as possible and from several out-of-state and private collections as well. Using the U.S. System of Rectangular Surveys (townships/sections grid system) to plot location data, he manually placed colored pencil dots representing about 21,000 specimens belonging to twenty-seven institutional and two private herbaria onto blank working maps, mapping only specimens that had been annotated by James H. Zimmerman or himself or occasionally another authority on the genus. The many specimen labels that do not include such information were mapped to the nearest named place. The computer-generated maps were produced from the same survey data and/or place information. Employing the Landnet file from the GEODISC2.1 produced by the Wisconsin Department of Natural Resources, Geographic Services Section, the x,y coordinate was determined, then plotted by ArcView graphic information system software. A comparison of the computer-generated maps with the unpublished floristic atlas maps revealed that many collections that had been mapped manually were yet to be databased. Therefore, the database maps that are the basis of this appendix have been updated through the addition of county records copied from the manuscript maps.

The databasing of the Wisconsin vascular flora has occupied the attention of the staff of the Wisconsin State Herbarium since 1994. Grants from the Wisconsin Department of Natural Resources and National Science Foundation and funds from the Department of Botany and O. N. and E. K. Allen Herbarium Fund, of the University of Wisconsin–Madison, have enabled us to enter label data from WIS, a task nearing completion for all Wisconsin vascular plants in the holdings. For the privilege of incorporating records from collections that have already been databased we extend our most sincere thanks to the staffs of the Milwaukee Public Museum, the Morton Arboretum, and the University of Wisconsin–Green Bay.

1. *Carex jamesii* Schweinitz

2. *Carex backii* W. Boott

3. *Carex pauciflora* Lightfoot

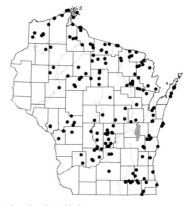

4. *Carex leptalea* Wahlenberg

5. *Carex hirtifolia* Mackenzie

6. *Carex pedunculata* Muhlenberg ex Willdenow

7. *Carex concinna* R. Brown

8. *Carex richardsonii* R. Brown

9. *Carex deflexa* Hornemann

10. *Carex umbellata* Schkuhr ex Willdenow

11. *Carex tonsa* (Fernald) E. P. Bicknell

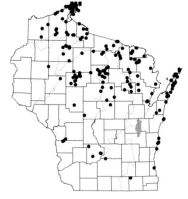

12. *Carex communis* L. H. Bailey

13. *Carex lucorum* Willdenow ex Link

14. *Carex inops* L. H. Bailey

15. *Carex pensylvanica* Lamarck

16. *Carex peckii* Howe

17. *Carex novae-angliae* Schweinitz

18. *Carex albicans* Willdenow ex Sprengel

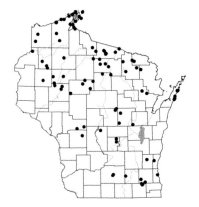

19. *Carex limosa* Linnaeus

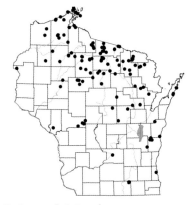

20. *Carex magellanica* Lamarck

21. *Carex swanii* (Fernald) Mackenzie

22. *Carex bushii* Mackenzie

23. *Carex pallescens* Linnaeus

24. *Carex torreyi* Tuckerman

25. *Carex scabrata* Schweinitz

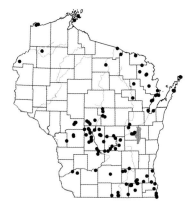

26. *Carex buxbaumii* Wahlenberg

27. *Carex media* R. Brown ex Richardson

28. *Carex eburnea* W. Boott

29. *Carex vaginata* Tauscher

30. *Carex woodii* Dewey

31. *Carex livida* (Wahlenberg) Willdenow

32. *Carex meadii* Dewey

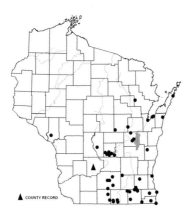

33. *Carex tetanica* Schkuhr

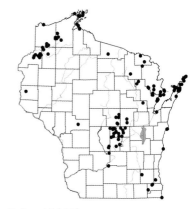

34. *Carex viridula* Michaux

35. *Carex cryptolepis* Mackenzie

36. *Carex flava* Linnaeus

37. *Carex prasina* Wahlenberg

38. *Carex gracillima* Schweinitz

39. *Carex davisii* Schweinitz & Torrey

40. *Carex formosa* Dewey

41. *Carex sprengelii* Dewey ex Sprengel

42. *Carex capillaris* Linnaeus

43. *Carex assiniboinensis* W. Boott

44. *Carex castanea* Wahlenberg

45. *Carex arctata* W. Boott

46. *Carex debilis* Michaux

47. *Carex albursina* E. Sheldon

48. *Carex gracilescens* Steudel

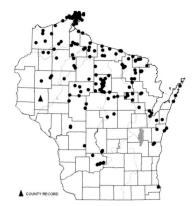

49. *Carex leptonervia* (Fernald) Fernald

50. *Carex blanda* Dewey

51. *Carex ormostachya* Wiegand

52. *Carex laxiflora* Lamarck

53. *Carex crawei* Dewey ex Torrey

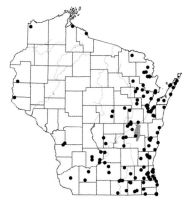

54. *Carex granularis* Muhlenberg ex Willdenow

55. *Carex conoidea* Schkuhr ex Willdenow

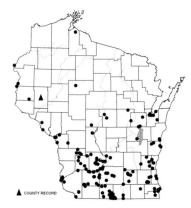

56. *Carex grisea* Wahlenberg

57. *Carex hitchcockiana* Dewey

58. *Carex oligocarpa* Schkuhr ex Willdenow

59. *Carex plantaginea* Lamarck

60. *Carex careyana* Torrey ex Dewey

61. *Carex platyphylla* J. Carey

62. *Carex laxiculmis* Schweinitz

63. *Carex digitalis* Willdenow

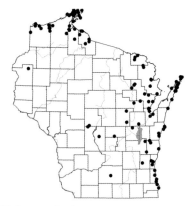

64. *Carex aurea* Nuttal

65. *Carex garberi* Fernald

66. *Carex crinita* Lamarck

67. *Carex gynandra* Schweinitz

68. *Carex aquatilis* Wahlenberg

69. *Carex lenticularis* Michaux

70. *Carex stricta* Lamarck

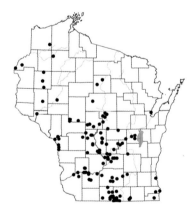

71. *Carex haydenii* Dewey

72. *Carex nigra* (Linnaeus) Reichard

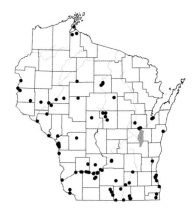

73. *Carex emoryi* Dewey ex Torrey

74. *Carex typhina* Michaux

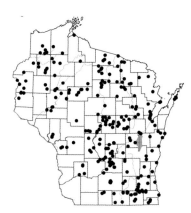

75. *Carex comosa* W. Boott

76. *Carex pseudocyperus* Linnaeus

77. *Carex schweinitzii* Dewey ex Schweinitz

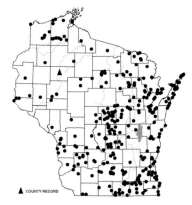

78. *Carex hystericina* Muhlenberg ex Willdenow

79. *Carex lurida* Wahlenberg

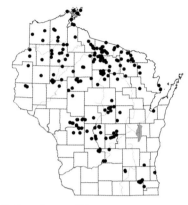

80. *Carex oligosperma* Michaux

81. *Carex tuckermanii* Dewey

82. *Carex retrorsa* Schweinitz

83. *Carex vesicaria* Linnaeus

84. *Carex utriculata* W. Boott

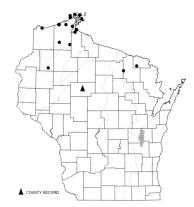

85. *Carex rostrata* Stokes

86. *Carex lacustris* Willdenow

87. *Carex houghtoniana* Torrey ex Dewey

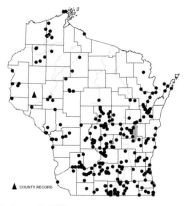

88. *Carex pellita* Willdenow

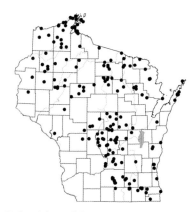

89. *Carex lasiocarpa* Ehrhart

90. *Carex trichocarpa* Muhlenberg ex Willdenow

91. *Carex hirta* Linnaeus

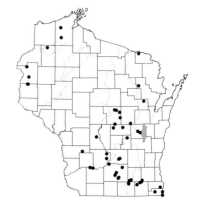

92. *Carex atherodes* Sprengel

93. *Carex laeviconica* Dewey

94. *Carex folliculata* Linnaeus

95. *Carex michauxiana* Boeckeler

96. *Carex intumescens* Rudge

97. *Carex grayi* J. Carey

98. *Carex lupulina* Muhlenberg ex Willdenow

99. *Carex lupuliformis* Sartwell ex Dewey

100. *Carex praegracilis* W. Boott

101. *Carex gynocrates* Wormskjöld ex Drejer

102. *Carex sartwellii* Dewey

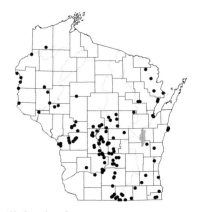

103. *Carex siccata* Dewey

104. *Carex vulpinoidea* Michaux

105. *Carex annectens* (E. P. Bicknell) E. P. Bicknell

106. *Carex diandra* Schrank

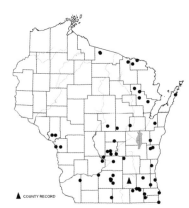

107. *Carex prairea* Dewey ex Wood

108. *Carex stipata* Muhlenberg ex Willdenow

109. *Carex crus-corvi* Shuttleworth ex G. Kunze

110. *Carex laevivaginata* (Kükenthal) Mackenzie

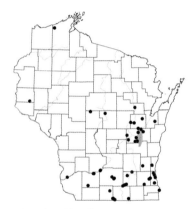

111. *Carex alopecoidea* Tuckerman

112. *Carex chordorrhiza* Ehrhart ex Linnaeus

113. *Carex sparganioides* Muhlenberg ex Willdenow

114. *Carex cephaloidea* (Dewey) Dewey

115. *Carex gravida* L. H. Bailey

116. *Carex aggregata* Mackenzie

117. *Carex rosea* Schkuhr ex Willdenow

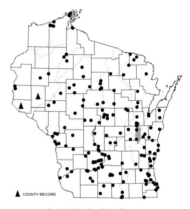

118. *Carex radiata* (Wahlenberg) Small

119. *Carex muehlenbergii* Schkuhr ex Willdenow

120. *Carex spicata* Hudson

121. *Carex cephalophora* Muhlenberg ex Willdenow

122. *Carex leavenworthii* Dewey

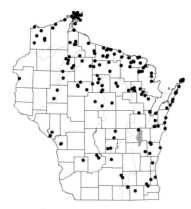

123. *Carex disperma* Dewey

124. *Carex tenuiflora* Wahlenberg

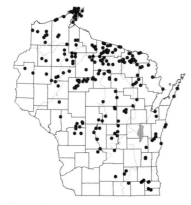

125. *Carex trisperma* Dewey

126. *Carex arcta* Boott

127. *Carex canescens* Linnaeus

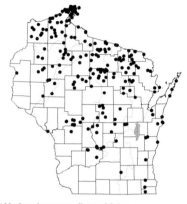

128. *Carex brunnescens* (Persoon) Poiret

129. *Carex deweyana* Schweinitz

130. *Carex bromoides* Schkuhr ex Willdenow

131. *Carex exilis* Dewey

132. *Carex sterilis* Willdenow

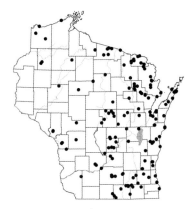

133. *Carex interior* L. H. Bailey

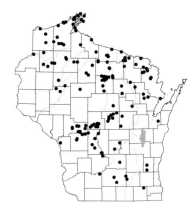

134. *Carex echinata* Murray

135. *Carex sychnocephala* J. Carey

136. *Carex muskingumensis* Schweinitz

137. *Carex cristatella* Britton

138. *Carex tribuloides* Wahlenberg

139. *Carex projecta* Mackenzie

140. *Carex adusta* Boott

141. *Carex foenea* Willdenow

142. *Carex ovalis* Goodenough

143. *Carex straminea* Willdenow ex Schkuhr

144. *Carex suberecta* (Olney) Britton

145. *Carex cumulata* (L. H. Bailey) Mackenzie

146. *Carex longii* Mackenzie

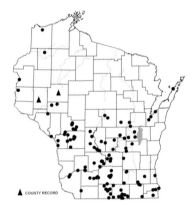

147. *Carex bicknellii* Britton

148. *Carex merritt-fernaldii* Mackenzie

149. *Carex molesta* Mackenzie ex Bright

150. *Carex festucacea* Schkuhr ex Willdenow

151. *Carex brevior* (Dewey) Mackenzie ex Lunell

152. *Carex crawfordii* Fernald

153. *Carex scoparia* Schkuhr ex Willdenow

154. *Carex bebbii* (L. H. Bailey) Olney ex Fernald

155. *Carex normalis* Mackenzie

156. *Carex tincta* (Fernald) Fernald

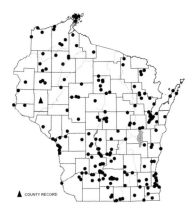

157. *Carex tenera* Dewey

Glossary

Abaxial. In a position facing away from the plant axis. The abaxial face of a leaf blade is the lower or outer face, unless the leaf is twisted. Cf. *adaxial.*

Achene. A dry, stonelike, indehiscent (not splitting open at maturity), one-seeded fruit in which the seed is not readily separated from the fruit wall; this is the fruit type throughout the sedge family. In most sedges other than *Carex* the achene surface is textured and highly variable from one species to the next. In *Carex* the achene surface is typically smooth.

Acuminate. Narrowing to a slender, acute tip with concave sides, the tip readily differentiated from the blade or body.

Adaxial. In a position facing the plant axis. The adaxial face of a leaf blade is the upper or inner face, unless the leaf is twisted. Cf. *abaxial.*

Adventive. Introduced and persisting, but not widely naturalized.

Androgynous. Bisexual, with staminate (male) flowers at the tip, pistillate (female) flowers at the base. Androgyny can be difficult to detect, as the staminate flowers are often inconspicuous, hidden among the tips of the ripe perigynia. For clear illustrations of androgynous spikes, see *Carex pedunculata* [6], *C. cryptolepis* [35], and *C. stricta* [70] and relatives.

Anther. The saclike, pollen-producing structure of a flower, typically borne on a filament; the fertile portion of the stamen. An abundance of yellow anthers is visible in the staminate spike illustration for *Carex richardsonii* [8].

Aphyllopodic. Having basal sheaths with reduced leaf blades (cataphylls). *Carex pedunculata* [6] is conspicuously aphyllopodic. Cf. *phyllopodic.*

Appressed. Angled upward at the base and lying parallel and close to the structure that it overlaps, as in roof shingles or fish scales. The perigynia in *Carex muskingumensis* [136] are appressed.

Ascending. Inclined upward (as in ascending culms) or toward the tip of the plant (as in ascending perigynia); differs from appressed by either spreading at the base or not lying closely parallel to the structure above. The perigynia of *Carex intumescens* [96] are ascending.

Awn. A slender bristle, typically arising from the apex or back of the structure that bears it. Numerous sedges have pistillate scales that bear awns, always from the apex.

Beak. Generally a slender and often somewhat elongated terminal appendage on a thickened organ such as a seed or fruit (not a flat organ such as a scale). In sedges, the shape and dimensions of the perigynium beak are often taxonomically useful.

Biconvex. Convex on both faces; lenticular. Cf. *planoconvex.*

Bidentate. Two-toothed. *Carex bicknellii* [147] has bidentate perigynium beaks.

Bifid. Cleft from the tip into two parts.

Bract. Vegetative, leaflike structure associated with the inflorescence, arising at the base of an individual spike or the entire inflorescence. Technically, scales are chaffy, dry, reduced bracts that subtend single flowers, but, for the sake of clarity, the two terms are treated in this book as though they were mutually exclusive. The lowest bract of *Carex aquatilis* [68] overtops the inflorescence, helping to set it apart from the other common species of its section.

Capillary. Hair-like.

Capitate. Shaped like a head, or gathered into a dense, head-like cluster. The inflorescences of *Carex bebbii* [154] and *C. cephalophora* [121] are capitate.

Cauline. Borne on or relating to the stem or culm.

Cespitose. Growing in discrete and often dense clumps: "densely cespitose" means having many shoots clumped together, "loosely cespitose" means having few shoots to a clump. Cespitose plants often have short rhizomes (e.g., *Carex blanda* [50]), but the cespitose habit is associated with long-creeping rhizomes in some species (e.g., *C. eburnea* [28]).

Cladoprophyll. A tubular bract that sheaths the base of the inflorescence stalk in subgenus *Carex*, typically hidden inside the bract sheath.

Clavate. Club-shaped, narrow at the base and thicker at the tip. The terminal spike in *Carex interior* [133] is clavate.

Corrugated. Rugose: having numerous horizontal, parallel ridges, creases, or folds, suggesting riffles on a dirt road or the textured surface of a washboard. The inner band of the leaf sheath in *Carex stipata* [108] is corrugated.

Culm. The stem of a grass, sedge, or rush. It is not improper to refer to sedge culms as stems, but "culm" is generally used out of tradition.

Deciduous. Separating and falling from the supporting structure while the latter is active, not yet dormant.

Decumbent. Partly reclining against the ground or other supporting surface at least toward the base, the endmost portion raised.

Dioecious. Unisexual, with all flowers on a given plant either pistillate or staminate.

Divergent. Pointing outward and at approximately right angles to the axis of attachment. The perigynia of *Carex hystericina* [78] and *C. grayi* [97] are divergent.

Drooping. Having the body arched so that the base orients upward and the tip orients downward. Some of the pistillate spikes of *Carex crinita* [66] and *C. comosa* [75] droop. Cf. *pendulous*.

Entire. Smooth, untoothed; used to describe the margin of a structure (e.g., a leaf, scale, or perigynium beak).

Epidermis. A plant's "skin"—the outermost layer of cells, adapted to protect the plant from desiccation and damage while allowing respiration and photosynthesis to take place. The epidermis on the inner face of the perigynium in *Carex bicknellii* [147] is translucent, revealing the dark achene.

Fibrillose. Covered with fibers or fiberlike structures. The bases of *Carex pensylvanica* [15] are clothed in fibers that run vertically, parallel to the plant axis, while the bases of *C. stricta* [70] are ladder-fibrillose or pinnate-fibrillose, clothed in fibers that run more or less horizontally and suggest the rungs of a ladder.

Filiform. Threadlike.

Foliose. Leaflike. The lower pistillate scales of *Carex jamesii* [1] are foliose.

Glabrous. Hairless, smooth. Cf. *pubescent, scabrous*.

Glaucous. Covered with a whitish, waxy substance that rubs off easily.

Globose. Approximately spherical; orbicular. Applies only to three-dimensional forms. The perigynium body of *Carex pensylvanica* [15] is approximately globose.

Graminoid. Grasses (Poaceae) and grasslike plants, such as sedges (Cyperaceae) and rushes (Juncaceae).

Gynecandrous. Bisexual, with pistillate (female) flowers at the tip, staminate (male) flowers at the base. For clear illustrations of gynecandrous spikes see *Carex buxbaumii* [26], *C. gracillima* [38], *C. interior* [133], *C. tenera* [157].

Hyaline. Translucent. The inner band of the leaf sheath in many *Carex* species is whitish-hyaline.

Impressed. Sunken, appearing to be inscribed into the surface on which it sits. The perigynium nerves of *Carex conoidea* [55] and *C. grisea* [56] are impressed.

Inflated. Having the appearance of being swollen, like a balloon; thin-walled and hollow, with convex surfaces. Perigynia of *Carex grayi* [97] and relatives are inflated.

Involute. Rolled toward the inner (adaxial) face.

Ladder-fibrillose. Clothed in a ladderlike reticulum of horizontal fibers. The basal leaf sheaths of *Carex stricta* [70] are ladder-fibrillose.

Lanceolate. Narrow, elongate, broadest below the middle and tapering to both ends. The perigynia of *Carex pauciflora* [3] are lanceolate.

Lenticular. Lens-shaped: thin, with two convex faces opposite one another, intersecting at a narrow edge, shaped like a lentil; biconvex. Cf. *planoconvex.*

Ligule. A membranous extension of the leaf sheath, originating at the point where the leaf sheath and leaf blade meet. In sedges the ligule is mostly fused to the inner face of the leaf blade, with a loose margin; in grasses the ligule is free except at the base, where it attaches to the leaf sheath, and may be fibrous or membranous.

Lustrous. Glossy, shiny.

Morphology. The study of forms. Within biology, morphologists typically concern themselves with the development and evolution of plant parts. A morphologist might study the perigynium, for instance, through developmental series and through comparison with inflorescence characters among the Cyperaceae that lack perigynia. The term *morphology* is often used as a synonym for the plant body itself, as in "describe the morphology of *Carex arcta.*"

Ob-. A prefix denoting that the shape it modifies is inverted (as though flipped upside-down). An ovate perigynium is roughly egg-shaped, widest below the middle; an obovate perigynium is roughly egg-shaped but widest above the middle. Any terms in this book that begin with ob- can be found under the root word.

Orbicular. Globose. Cf. *orbiculate.*

Orbiculate. Circular in outline; applies only to flat (plane) shapes and outlines. The perigynium in *Carex brevior* [151] is orbiculate. Cf. *orbicular.*

Ovate. Egg-shaped; broadest below the middle, round at the base, and tapering to a rounded apex. This term applies to two-dimensional structures or projections (e.g., the outline of a pistillate scale). Cf. *ovoid.*

Ovoid. Egg-shaped. This term applies to three-dimensional shapes. Cf. *ovate.*

Paniculate. Composed of panicles, branching inflorescence units. The compound-spiked subgenus *Vignea* sections (*Vulpinae, Multiflorae, Heleoglochin,* and *Phaestoglochin* in part) have paniculate inflorescences.

Papilla (pl.: papillae; adj.: papillose). A small, blisterlike protrusion of the epidermis. Papillae are best viewed with 10–30× magnification.

Peduncle. The stalk of an inflorescence or major (primary) inflorescence unit; used in *Carex* for the stalk of the individual spikes, which may be sessile or pedunculate (borne on a peduncle).

Pendulous. Hanging downward from a slender stalk. The lateral spikes of *Carex gracillima* [38] are pendulous.

Perigynium. A saclike structure that surrounds the pistillate flower in *Carex* and a handful of closely related sedge genera. In these latter genera the perigynium may be only partially closed, but in *Carex* it is entirely closed except for the orifice at the apex. The stigma protrudes from the tip of the perigynium, and the achene develops within. See utricle.

Phyllopodic. Describes plants in which the basal leaf sheaths of fertile culms bear well-developed blades. Phyllopody is a consequence of vegetative shoots becoming fertile; thus, the fertile culm arises from a rosette of older, senescent leaves. Cf. *aphyllopodic.*

Pistil. Carpel: the seed-bearing structure of the angiosperm flower, comprising the ovary, style, and stigma. Cf. *stamen.*

Pistillate. Bearing functional pistils but not stamens; functionally female. Cf. *staminate.*

Planoconvex. Flat on one face, convex on the other. Cf. *biconvex, lenticular.*

Pseudoculm. Vegetative shoots that may be elongated but do not include a true culm with nodes and internodes. A cross-section just above the base of a pseudoculm will reveal a series of leaf sheaths wrapped around each other with no stem tissue in the middle. Cf. *vegetative culm.*

Pubescent. Bearing trichomes (hairs). This is a very general term that encompasses a bewildering array of more precise terms that are largely ignored in this book for the sake of simplicity. Cf. *glabrous, scabrous.*

Reflexed. Abruptly bent downward or backward. The perigynia of *Carex pauciflora* [3] and *C. comosa* [75] are reflexed.

Revolute. Curled toward the outer (abaxial) face.

Rhizome. An underground stem, superficially rootlike but differentiated into nodes and internodes, often bearing fibers at the nodes, and lacking root hairs. Rhizomatous plants are often not strongly cespitose, though they can be. The very common *Carex pensylvanica* [15] and *C. siccata* [103] are conspicuously rhizomatous.

Rugose. See corrugated.

Scabrous. Rough; bearing short, stiff protrusions or hairs.

Scale. A dry, reduced bract; sedge flowers are each *subtended* by a single scale. The distinction between scales and bracts breaks down at times. The lowest pistillate scales of *Carex jamesii* [1], for instance, are elongate and bractlike (foliose), but they are referred to as scales because they subtend a single flower rather than a portion of the inflorescence. Pistillate scales in *C. limosa* [19] are very prominent, largely covering the perigynia.

Septate-nodulose. Textured (bumpy) due to the presence of septa that become distinct upon drying due to shrinkage of the surrounding tissue. The leaves of *Carex comosa* [75] are septate-nodulose. Cf. *septum.*

Septum (pl.: septa). A distinct wall that separates chambers in a leaf, fruit, or other organ. Cf. *septate-nodulose.*

Serrate (diminutive: serrulate). Finely toothed.

Sessile. Not borne on a stalk. Spikes in subgenus *Vignea* are typically sessile. Cf. *peduncle.*

Setaceous. Bristlelike: slender, stiff, elongate, narrowed to a slender tip. The inflorescence of *Carex cephalophora* [121] has numerous setaceous bracts. Cf. *foliose.*

Spike. Generally, an elongate inflorescence bearing sessile flowers. In the Cyperaceae the spike is the unit of the inflorescence that bears sessile flowers subtended by their respective scales. An entire inflorescence is typically composed of several spikes (with the exception of unispicate species such as *Carex jamesii* [1], *C. pauciflora* [3], and *C. leptalea* [4]).

Stamen. The pollen-producing organ of an angiosperm flower, typically comprising a slender filament bearing a saclike structure called the anther. Cf. *pistil.*

Staminate. Bearing functional stamens but not pistils; functionally male. Cf. *pistillate.*

Stigma. The endmost portion of the pistil, which intercepts pollen and stimulates the pollen tube to grow. The stigma in sedges is typically forked into two or three segments (bifid or trifid, respectively). Indivduals with bifid stigmas are often referred to as distigmatic, that is, having two stigmas.

Stipe. A diminutive stalk; most typically applied to such structures at the bases of achenes and perigynia. The perigynium in *Carex arctata* [45] is stipitate.

Stolon. A stem that grows horizontally above or just beneath the ground surface, rooting at the nodes and producing shoots at some nodes and/or at the apex. The distinction between stolon and rhizome is not always clear, but typically a stolon is more shootlike in appearance than a rhizome. *Carex chordorrhiza* [112] is stoloniferous.

Style. The slender, stalk-like structure that connects the stigma to the ovary.

Subtend. Emerge immediately beneath the subtended organ. Pistillate scales subtend the perigynia immediately above them.

Subterete. See terete.

Terete. Round in cross-section, longer than wide; a baseball bat is terete, but a baseball is not. Subterete: approximately terete.

Tomentum (adj.: tomentose). Dense, woolly pubescence. The roots of *Carex limosa* [19] and allies are covered with a yellow, feltlike tomentum.

Trigonous. Three-angled.

Truncate. Blunt, appearing as if the apex had been cut off.

Unispicate. Bearing a single spike. *Carex jamesii* [1], *C. pauciflora* [3], and *C. leptalea* [4] are unispicate.

Utricle. A small, bladder-shaped organ. The term is sometimes used in reference to the sac that surrounds the pistillate flower in *Carex* in lieu of the more specific term perigynium (*q.v.*).

Vegetative culm. A vegetative shoot with a true culm, which is differentiated into nodes and internodes. A cross-section through a vegetative culm will reveal an envelope of leaf sheath(s) surrounding a pithy or hollow culm. *Carex muskingumensis* [136] and *C. tribuloides* [138] have prominent vegetative culms. Cf. *pseudoculm*.

Bibliography

Ball, P. W., and A. A. Reznicek, eds. 2002. *Carex* Linnaeus. In *Flora of North America*, vol. 23: *Magnoliophyta: Commelinidae (in part): Cyperaceae,* ed. Flora of North America Editorial Committee, 254–573. New York: Oxford University Press. This volume is the most current, most definitive guide to the sedges of North America. It contains keys and descriptions for all taxa in the flora recognized at the time of writing as well as descriptions of and some commentary on the North American sections of *Carex.* Written by approximately twenty different taxonomists, the *Carex* treatment represents a wide range of expertise. The contents of the volume—including keys, maps, and illustrations—are available online at http://www.efloras.org/florataxon.aspx?flora_id=1&taxon_id=10246.

Cronquist, A. 1988. *The evolution and classification of flowering plants.* Bronx: New York Botanical Garden.

Curtis, J. T. 1959. *The vegetation of Wisconsin.* Madison: University of Wisconsin Press.

Deam, C. C. 1940. *Flora of Indiana.* Indianapolis: Indiana Department of Conservation, Division of Forestry.

Fassett, N. C. 1976. *Spring flora of Wisconsin.* 4th ed. Madison: University of Wisconsin Press. The sedge treatment in this book was written by the late James H. Zimmerman, with illustrations by Elizabeth H. Zimmerman. The keys are straightforward and easy to use, but some less common species are omitted. However, the omission of less frequently encountered taxa makes the keys somewhat easier to use, and this book is a useful reference, especially if you are just learning Wisconsin's sedges.

Fernald, M. L. 1950. *Gray's manual of botany.* 8th ed. New York: American Book Company. This is an older book, but Fernald's keys and descriptions are detailed and readable. If you are serious about learning the flora of the Northeast and Midwest, get your hands on this book.

Galatowitsch, S. M., and A. G. van der Valk. 1996. Vegetation and environmental conditions in recently restored wetlands in the prairie pothole region of the USA. *Vegetatio* 126:89–99. This paper demonstrates that restoring hydrology is insufficient to get sedges back into degraded wetlands.

Gleason, H. A., and A. Cronquist. 1991. *Manual of vascular plants of northeastern United States and adjacent Canada.* 2nd ed. New York: New York Botanical Garden. A great book and in many ways practical. The sedge treatment is difficult to use and not recommended for beginners. The illustrations in the companion volume (Holmgren and Holmgren 1998) are helpful.

Handel, S. N. 1976. Dispersal ecology of *Carex pedunculata* (Cyperaceae), a new North American myrmecochore. *American Journal of Botany* 63:1071–79.

Hipp, A. L. 1998. A checklist of carices for prairies, savannas and oak woodlands of southern Wisconsin. *Transactions of the Wisconsin Academy of Sciences, Arts and Letters* 86:77–99.

Holmgren, N. H., and P. K. Holmgren, eds. 1998. *Illustrated companion to Gleason and Cronquist's manual.* New York: New York Botanical Garden. See comments under Gleason and Cronquist 1991.

Hujik, P. M. 1995. Lowland savannas: Groundlayer composition and distribution in relation to elevation and light. M.S. thesis, University of Wisconsin–Madison.

Kettenring, K. M. 2006. Seed ecology of wetland Carex spp.: Implications for restoration. Ph.D. dissertation, University of Minnesota, St. Paul.

Kirschbaum, C. D. 2007. The taxonomy of *Carex* trisperma (Cyperaceae). *Journal of the Botanical Research Institute of Texas* 1:389–405.

Kükenthal, G. 1909. Cyperaceae—Caricoideae. In *Das Pflanzenreich IV*, ed. A. Engler, 1–824. Leipzig: Wilhelm Engelmann. The only worldwide monograph for the genus *Carex.*

Mackenzie, K. K. 1935. *Carex:* Cariceae. *North American Flora* 18:169–478. The single most important publication in twentieth-century North American caricology, covering every species known in North America at the time of publication. The accompanying illustrations by H. C. Creutzberg are magnificent, but they were published in a separate, oversized volume (K. K. Mackenzie, 1940, *North American Cariceae* [New York: New York Botanical Garden]) and are only available in relatively specialized libraries. The plates have been digitized and are available online through the Texas A&M University Bioinformatics Working Group website: http://www.csdl.tamu.edu/FLORA/carex/carexout.htm.

Mohlenbrock, R. H. 1999. *The illustrated flora of Illinois—sedges:* Carex. Carbondale: Southern Illinois University Press. This book provides keys, detailed descriptions, and illustrations for all *Carex* species in Illinois. The illustrations alone are worth the price of the book.

Reznicek, A. A. 1990. Evolution in sedges (*Carex,* Cyperaceae). *Canadian Journal of Botany* 68:1409–32. This rather technical article is the best discussion of morphological patterns of evolution within *Carex.* The article is part of the proceedings of the 1987 symposium "Systematics and Ecology of the Genus *Carex* (Cyperaceae)" held in Montreal. Other articles in the volume cover topics such as phytogeography, vegetative reproduction, anatomy, and taxonomic inference based on morphological and chemical/molecular traits.

Robert W. Freckmann Herbarium Website. http://wisplants.uwsp.edu. A great resource for the Wisconsin flora, including interactive maps that link to herbarium specimens. Nomenclature follows Wetter et al. 2001.

Rothrock, P. E., and A. A. Reznicek. 2000. Taxonomy, ecology, and biogeography of *Carex* section *Ovales* in Indiana. *Michigan Botanist* 39:19–37. This paper covers sixteen of the twenty-three-section *Ovales* species in the Wisconsin flora, including much interesting information on habitats and identification.

Schütz, W. 2000. Ecology of seed dormancy and germination in sedges (*Carex*). *Perspectives in Plant Ecology, Evolution, and Systematics* 3:67–89.

Swink, F., and G. Wilhelm. 1994. *Plants of the Chicago region.* 4th ed. Indianapolis: Indiana Academy of Science. This book is an excellent source of information about habitats of all vascular plant species (including sedges) in the twenty-two-county Chicago region. Keys to all species are provided, but species descriptions are not. The book is out of print, but it is being revised, and the fourth edition is widely available in libraries. The list of taxa in the book is largely the same as that used by vPlants.

van der Valk, A. G., T. L. Bremholm, and E. Gordon. 1999. The restoration of sedge meadows: Seed viability, seed germination requirements, and seedling growth of *Carex* species. *Wetlands* 19:756–64.

Voss, E. G. 1972. *Michigan flora,* vol. 1: *Gymnosperms and monocots.* Ann Arbor: Cranbrook Institute of Science and University of Michigan Herbarium. Dr. Voss's keys throughout this volume work beautifully, and the illustrations and selective information on habitats and identification strike a fine balance between the needs of the taxonomist and the interests of the naturalist.

vPlants: A virtual herbarium of the Chicago region. http://www.vplants.org. An online, searchable database of Chicago region herbarium specimens and information on plant identification. The addition of an online field guide provides additional identification for Midwest sedges and other plants.

Werner, K. J., and J. B. Zedler. 2002. How sedge meadow soils, microtopography, and vegetation respond to sedimentation. *Wetlands* 22:451–466. This paper provides strong evidence that *Carex stricta,* by forming peaty tussocks, creates microsites for a wide range of sedge meadow species. Thus, tussock sedge directly increases the biodiversity of wetlands, but this effect is reversed when tussocks are buried beneath heavy, mineral-rich sediment.

Wetter, M. A., T. S. Cochrane, M. R. Black, H. H. Iltis, and P. E. Berry. 2001. *Checklist of the vascular plants of Wisconsin.* DNR Technical Bulletin 192. Department of Natural Resources, Madison. The authoritative list of plants known in Wisconsin. A searchable, updated version is available online with links to maps, photos, and additional information at http://www.botany.wisc.edu/wisflora/.

Wheeler, G. A., and G. B. Ownbey. 1984. Annotated list of Minnesota carices, with phyto-geographical and ecological notes. *Rhodora* 86:151–231.

Wisconsin Department of Natural Resources. January 2004. Wisconsin Natural Heritage Working List. Department of Natural Resources, Bureau of Endangered Resources, Madison.

Taxonomic Index

This index includes all scientific names mentioned in a taxonomic context, including discussion of similar species. References to associated species are not included in this list. Taxonomic names are in roman-face if they are accepted names, in italics if they are synonyms. Bold-faced type indicates pages on which a taxon is described in detail, either as the main species being described or as a similar species with an illustration. Maps are indicated in italicized bold-face.